T0178550

# Communications
# in Computer and Information Science 1991

## Rationale

The CCIS series is devoted to the publication of proceedings of computer science conferences. Its aim is to efficiently disseminate original research results in informatics in printed and electronic form. While the focus is on publication of peer-reviewed full papers presenting mature work, inclusion of reviewed short papers reporting on work in progress is welcome, too. Besides globally relevant meetings with internationally representative program committees guaranteeing a strict peer-reviewing and paper selection process, conferences run by societies or of high regional or national relevance are also considered for publication.

## Topics

The topical scope of CCIS spans the entire spectrum of informatics ranging from foundational topics in the theory of computing to information and communications science and technology and a broad variety of interdisciplinary application fields.

## Information for Volume Editors and Authors

Publication in CCIS is free of charge. No royalties are paid, however, we offer registered conference participants temporary free access to the online version of the conference proceedings on SpringerLink (http://link.springer.com) by means of an http referrer from the conference website and/or a number of complimentary printed copies, as specified in the official acceptance email of the event.

CCIS proceedings can be published in time for distribution at conferences or as post-proceedings, and delivered in the form of printed books and/or electronically as USBs and/or e-content licenses for accessing proceedings at SpringerLink. Furthermore, CCIS proceedings are included in the CCIS electronic book series hosted in the SpringerLink digital library at http://link.springer.com/bookseries/7899. Conferences publishing in CCIS are allowed to use Online Conference Service (OCS) for managing the whole proceedings lifecycle (from submission and reviewing to preparing for publication) free of charge.

## Publication process

The language of publication is exclusively English. Authors publishing in CCIS have to sign the Springer CCIS copyright transfer form, however, they are free to use their material published in CCIS for substantially changed, more elaborate subsequent publications elsewhere. For the preparation of the camera-ready papers/files, authors have to strictly adhere to the Springer CCIS Authors' Instructions and are strongly encouraged to use the CCIS LaTeX style files or templates.

## Abstracting/Indexing

CCIS is abstracted/indexed in DBLP, Google Scholar, EI-Compendex, Mathematical Reviews, SCImago, Scopus. CCIS volumes are also submitted for the inclusion in ISI Proceedings.

## How to start

To start the evaluation of your proposal for inclusion in the CCIS series, please send an e-mail to ccis@springer.com.

Marija Mihova · Mile Jovanov

Editors

# ICT Innovations 2023

## Learning: Humans, Theory, Machines, and Data

15th International Conference, ICT Innovations 2023
Ohrid, North Macedonia, September 24–26, 2023
Proceedings

Springer

*Editors*
Marija Mihova ⓘD
Ss. Cyril and Methodius University in Skopje
Skopje, North Macedonia

Mile Jovanov ⓘD
Ss. Cyril and Methodius University in Skopje
Skopje, North Macedonia

ISSN 1865-0929          ISSN 1865-0937 (electronic)
Communications in Computer and Information Science
ISBN 978-3-031-54320-3          ISBN 978-3-031-54321-0 (eBook)
https://doi.org/10.1007/978-3-031-54321-0

This Springer imprint is published by the registered company Springer Nature Switzerland AG
The registered company address is: Gewerbestrasse 11, 6330 Cham, Switzerland

Paper in this product is recyclable.

# Preface

This volume constitutes the refereed proceedings with selected papers and keynote talk abstracts of the 15th International Conference on ICT Innovations 2023, held in Ohrid, Republic of North Macedonia, during September 24–26, 2023. After several years of online and hybrid events due to the restrictions caused by the COVID-19 pandemic, this edition was onsite, uniting more than 100 researchers and participants who joined the conference and the satellite workshops.

The ICT Innovations conference series, led by the Macedonian Society of Information and Communication Technologies (ICT-ACT) and supported by the Faculty of Computer Science and Engineering (FCSE) in Skopje, has firmly established itself as a platform for presenting scientific findings in the realm of pioneering fundamental and applied research within ICT. It's a place to share the newest discoveries and practical solutions in ICT research, and to discuss the latest trends, opportunities, and challenges in computer science and engineering.

The central focus of this year's conference revolved around the theme of "Learning: Humans, Theory, Machines, and Data". In the post-COVID era, our conference explores the transformative journey of AI, machine learning, and human education heavily influenced by ICT. Focusing on the interplay of human and machine learning, we delve into how insights from human learning theory can enhance machine learning and vice versa. Emphasizing the role of data, we support using data-driven approaches to understand human-machine learning dynamics.

The key question guiding discussions is: Can machine learning models effectively learn from humans in a natural way, and reciprocally, can machine learning interpret human learning for improved outcomes? This practical exploration aims to use ICT's power for optimized learning, fostering collaboration across disciplines. In parallel with the evolving landscape of data science, our conference delves into the connection between human learning, machine learning, and rich data. Employing advanced analytics, we aim to push diverse fields forward, imagining a future where our grasp of learning reshapes education, technology, and society.

17 full papers included in this volume were carefully reviewed and selected from 52 submissions. They were organized by topic as follows: theoretical informatics; artificial intelligence and natural language processing; image processing; e-education and e-services; network science and medical informatics/bioinformatics. The conference covered a spectrum of topics, including digital transformation technologies and a broad range of subjects within computer science and engineering. The papers were thoroughly examined and graded by 87 reviewers from 33 countries, assigning for each paper three to five reviewers, with single-blind reviews. These reviewers were chosen based on their scientific excellence in specific domains, ensuring a rigorous and comprehensive review process.

Four keynote speakers prepared talks in line with the topic. We were privileged that our distinguished keynote speakers accepted the invitation and gave the impressive talks:

Synthetic Networks (Gesine Reinert, University of Oxford, UK); Why dependent types matter? (Thorsten Altenkirch, University of Nottingham, UK); Machine learning with guarantees (Aleksandar Bojchevski, University of Cologne, Germany); and Why is AI a social problem in 2023? (Marija Slavkovikj, University of Bergen, Norway).

Moreover, in addition to three workshops, the conference also offered exciting social activities to enhance the connections among the participants, and especially after a long time of isolation we are proud of this achievement.

At the end of the conference, a best paper prize was awarded. A joint jury consisting of program and scientific committee members made the decision to recognize the paper: Extracting Entities and Relations in Analyst Stock Ratings News.

We are grateful to all authors who passionately contributed to this collection, the reviewers who diligently evaluated the submissions, and the conference participants who enriched the event with their presence and discussions. Moreover, we cordially thank the FCSE computing center staff, who created and maintained an impeccable working environment during the conference. Additionally, we express our sincere appreciation to all who have been part of this journey, and we look forward to continued exploration and discovery in the years to come. Finally, a special note of heartfelt thanks to Ilinka Ivanoska for her exceptional dedication and invaluable assistance throughout the entire process, ensuring the success of our conference.

We look forward to seeing you at the following 16th edition of the conference next year.

Sincerely,

December 2023

Marija Mihova
Mile Jovanov

# Organization

## General Chairs

Marija Mihova        Ss. Cyril and Methodius University in Skopje, North Macedonia

Mile Jovanov        Ss. Cyril and Methodius University in Skopje, North Macedonia

## Program Commitee Chairs

Marija Mihova        Ss. Cyril and Methodius University in Skopje, North Macedonia

Mile Jovanov        Ss. Cyril and Methodius University in Skopje, North Macedonia

Ilinka Ivanoska        Ss. Cyril and Methodius University in Skopje, North Macedonia

Sasho Gramatikov        Ss. Cyril and Methodius University in Skopje, North Macedonia

## Program Committee

Aleksandra Dedinec        Ss. Cyril and Methodius University in Skopje, North Macedonia

Aleksandra Mileva        Goce Delčev University of Štip, North Macedonia

Aleksandra P. Mitrovikj        Ss. Cyril and Methodius University in Skopje, North Macedonia

Amelia Badica        University of Craiova, Romania

Amjad Gawanmeh        University of Dubai, UAE

Ana Madevska Bogdanova        Ss. Cyril and Methodius University in Skopje, North Macedonia

Andrea Kulakov        Ss. Cyril and Methodius University in Skopje, North Macedonia

Andrej Brodnik        University of Ljubljana, Slovenia

Andrej Grguric        Ericsson Nikola Tesla d.d., Croatia

Andreja Naumoski        Ss. Cyril and Methodius University in Skopje, North Macedonia

Antonio De Nicola        ENEA, Italy

| Antun Balaz | Institute of Physics Belgrade, Serbia |
| Arianit Kurti | Linnaeus University, Sweden |
| Betim Cico | EPOKA University, Albania |
| Biljana Mileva Boshkoska | Jožef Stefan Institute, Slovenia |
| Biljana Stojkoska | Ss. Cyril and Methodius University in Skopje, North Macedonia |
| Biljana Tojtovska | Ss. Cyril and Methodius University in Skopje, North Macedonia |
| Blagoj Ristevski | St. Kliment Ohridski University, North Macedonia |
| Bojan Ilijoski | Ss. Cyril and Methodius University in Skopje, North Macedonia |
| Bojana Koteska | Ss. Cyril and Methodius University in Skopje, North Macedonia |
| Boris Delibashic | University of Belgrade, Serbia |
| Christophe Trefois | University of Luxembourg, Luxembourg |
| David Guralnick | International E-Learning Association, USA |
| David Shafranek | Masaryk University, Czech Republic |
| Dejan Gjorgjevikj | Ss. Cyril and Methodius University in Skopje, North Macedonia |
| Dejan Spasov | Ss. Cyril and Methodius University in Skopje, North Macedonia |
| Denis Trcek | University of Ljubljana, Slovenia |
| Dilip Patel | London South Bank University, UK |
| Dimitar Trajanov | Ss. Cyril and Methodius University in Skopje, North Macedonia |
| Dragi Kocev | Jožef Stefan Institute, Slovenia |
| Eftim Zdravevski | Ss. Cyril and Methodius University in Skopje, North Macedonia |
| Elena Vlahu-Gjorgievska | University of Wollongong, Australia |
| Elinda Kajo Mece | Polytechnic University of Tirana, Albania |
| Emmanuel Conchon | University of Limoges, XLIM, UMR CNRS 7252, France |
| Federico Pittino | Silicon Austria Labs, Austria |
| Fisnik Dalipi | Linnaeus University, Sweden |
| Florin Pop | University Politehnica of Bucharest/National Institute for Research and Development in Informatics (ICI), Bucharest, Romania |
| Francesc Burrull | Universidad Politécnica de Cartagena, Spain |
| Fu-Shiung Hsieh | Chaoyang University of Technology, Taiwan |
| Georgina Mirceva | Ss. Cyril and Methodius University in Skopje, North Macedonia |

| | |
|---|---|
| Gjorgji Madjarov | Ss. Cyril and Methodius University in Skopje, North Macedonia |
| Gjorgji Strezoski | University of Amsterdam, The Netherlands |
| Goce Armenski | Ss. Cyril and Methodius University in Skopje, North Macedonia |
| Hieu Trung Huynh | Industrial University of Ho Chi Minh City, Vietnam |
| Hrachya Astsatryan | Institute for Informatics and Automation Problems, National Academy of Sciences of Armenia, Armenia |
| Hristijan Gjoreski | Ss. Cyril and Methodius University in Skopje, North Macedonia |
| Hristina Mihajloska | Ss. Cyril and Methodius University in Skopje, North Macedonia |
| Igor Ljubi | University of Zagreb, Croatia |
| Ilche Georgievski | University of Stuttgart, Germany |
| Ilinka Ivanoska | Ss. Cyril and Methodius University in Skopje, North Macedonia |
| Irina Mocanu | University Politehnica of Bucharest, Romania |
| Ivan Chorbev | Ss. Cyril and Methodius University in Skopje, North Macedonia |
| Ivan Kitanovski | Ss. Cyril and Methodius University in Skopje, North Macedonia |
| Ivica Dimitrovski | Ss. Cyril and Methodius University in Skopje, North Macedonia |
| Jatinderkumar Saini | Narmada College of Computer Application, India |
| John Gialelis | University of Patras, Greece |
| Josep Silva | Universitat Politècnica de València, Spain |
| Kalinka Kaloyanova | University of Sofia, Bulgaria |
| Katarina Trojacanec Dineva | Ss. Cyril and Methodius University in Skopje, North Macedonia |
| Katerina Zdravkova | Ss. Cyril and Methodius University in Skopje, North Macedonia |
| Kire Trivodaliev | Ss. Cyril and Methodius University in Skopje, North Macedonia |
| Kosta Mitreski | Ss. Cyril and Methodius University in Skopje, North Macedonia |
| Kostadin Mishev | Ss. Cyril and Methodius University in Skopje, North Macedonia |
| Ladislav Huraj | University of SS. Cyril and Methodius in Trnava, Slovakia |
| Lasko Basnarkov | Ss. Cyril and Methodius University in Skopje, North Macedonia |

| | |
|---|---|
| Ljiljana Trajkovic | Simon Fraser University, Canada |
| Ljupcho Antovski | Ss. Cyril and Methodius University in Skopje, North Macedonia |
| Loren Schwiebert | Wayne State University, USA |
| Luis Alvarez Sabucedo | Universidade de Vigo, Spain |
| Magdalena Kostoska | Ss. Cyril and Methodius University in Skopje, North Macedonia |
| Magdalena Punceva | Rutgers University, USA |
| Manjeet Rege | University of St. Thomas, USA |
| Manuel Silva | ISEP-IPP and INESC TEC CRIIS, Portugal |
| Mahouton N. Hounkonnou | University of Abomey-Calavi, Benin |
| Marcin Michalak | Silesian University of Technology, Poland |
| Marco Aiello | University of Stuttgart, Germany |
| Marco Porta | University of Pavia, Italy |
| Marija Mihova | Ss. Cyril and Methodius University in Skopje, North Macedonia |
| Marjan Gusev | Ss. Cyril and Methodius University in Skopje, North Macedonia |
| Martin Drlik | Constantine the Philosopher University in Nitra, Slovakia |
| Martin Gjoreski | Universit della Svizzera italiana, Italy |
| Massimiliano Zanin | IFISC (CSIC-UIB), Spain |
| Matus Pleva | Technical University of Koice, Slovakia |
| Melanija Mitrovic | University of Niš, Serbia |
| Mile Jovanov | Ss. Cyril and Methodius University in Skopje, North Macedonia |
| Milena Djukanovic | Univerzitet Crne Gore, ME |
| Milos Stojanovic | Visoka tehnička škola Niš, Serbia |
| Milos Jovanovik | Ss. Cyril and Methodius University in Skopje, North Macedonia |
| Mirjana Ivanovic | University of Novi Sad, Serbia |
| Miroslav Mirchev | Ss. Cyril and Methodius University in Skopje, North Macedonia |
| Monika Simjanoska | Ss. Cyril and Methodius University in Skopje, North Macedonia |
| Natasha Ilievska | Ss. Cyril and Methodius University in Skopje, North Macedonia |
| Natasha Stojkovikj | Goce Delčev University of Štip, North Macedonia |
| Nevena Ackovska | Ss. Cyril and Methodius University in Skopje, North Macedonia |
| Novica Nosovic | University of Sarajevo, Bosnia Herzegovina |
| Nikola Simidjievski | University of Cambridge, UK |

| | |
|---|---|
| Panche Ribarski | Ss. Cyril and Methodius University in Skopje, North Macedonia |
| Pece Mitrevski | St. Kliment Ohridski University, North Macedonia |
| Petre Lameski | Ss. Cyril and Methodius University in Skopje, North Macedonia |
| Petar Sokoloski | Ss. Cyril and Methodius University in Skopje, North Macedonia |
| Riste Stojanov | Ss. Cyril and Methodius University in Skopje, North Macedonia |
| Robertas Damasevicius | Silesian University of Technology, Poland |
| Rossitza Goleva | New Bulgarian University, Bulgaria |
| Sasho Gramatikov | Ss. Cyril and Methodius University in Skopje, North Macedonia |
| Sasko Ristov | University of Innsbruck, Austria |
| Sergio Ilarri | University of Zaragoza, Spain |
| Shuxiang Xu | University of Tasmania, Australia |
| Simona Samardjiska | Radboud University, The Netherlands |
| Silva Ghilezan | University of Novi Sad, Mathematical Institute SASA, Serbia |
| Slobodan Bojanic | Universidad Politécnica de Madrid, Spain |
| Slobodan Kalajdziski | Ss. Cyril and Methodius University in Skopje, North Macedonia |
| Smile Markovski | Ss. Cyril and Methodius University in Skopje, North Macedonia |
| Smilka Janeska-Sarkanjac | Ss. Cyril and Methodius University in Skopje, North Macedonia |
| Snezana Savoska | St. Kliment Ohridski University, North Macedonia |
| Stanimir Stoyanov | University of Plovdiv Paisii Hilendarski, Bulgaria |
| Sonja Gievska | Ss. Cyril and Methodius University in Skopje, North Macedonia |
| Suzana Loshkovska | Ss. Cyril and Methodius University in Skopje, North Macedonia |
| Tommy Boshkovski | Polytechnique Montréal, Canada, QMENTA Inc., USA |
| Ustijana Rechkoska-Shikoska | UIST Ohrid, North Macedonia |
| Vacius Jusas | Kaunas University of Technology, Lithuania |
| Vassil Grozdanov | SWU, Bulgaria |
| Verica Bakeva | Ss. Cyril and Methodius University in Skopje, North Macedonia |
| Vesna Dimitrova | Ss. Cyril and Methodius University in Skopje, North Macedonia |

| | |
|---|---|
| Vesna Dimitrievska | Silicon Austria Labs, Austria |
| Vesna Dimitrievska Ristovska | Ss. Cyril and Methodius University in Skopje, North Macedonia |
| Vladimr Sildi | Matej Bel University, Slovakia |
| Vladimir Trajkovik | Ss. Cyril and Methodius University in Skopje, North Macedonia |
| Xiangyan Zeng | Fort Valley State University, USA |
| Yoram Haddad | Jerusalem College of Technology, Israel |
| Zaneta Popeska | Ss. Cyril and Methodius University in Skopje, North Macedonia |
| Zlatko Varbanov | Veliko Tarnovo University, Bulgaria |

## Technical Committee

| | |
|---|---|
| Ilinka Ivanoska | Ss. Cyril and Methodius University in Skopje, North Macedonia |
| Ana Todorovska | Ss. Cyril and Methodius University in Skopje, North Macedonia |
| Dimitar Kitanovski | Ss. Cyril and Methodius University in Skopje, North Macedonia |
| Vlatko Spasev | Ss. Cyril and Methodius University in Skopje, North Macedonia |
| Jovana Dobreva | Ss. Cyril and Methodius University in Skopje, North Macedonia |

# Keynote Talks (Abstracts)

# Synthetic Networks

Gesine Reinert

University of Oxford, UK
https://scholar.google.com/citations?hl=en&user=2gvyN5oAAAAJ

**Abstract.** Synthetic data are increasingly used in computational statistics and machine learning. Some applications relate to privacy concerns, to data augmentation, and to method development. Synthetic data should reflect the underlying distribution of the real data, being faithful but also showing some variability. In this talk we focus on networks as a data type, such as networks of financial transactions. This data type poses additional challenges due to the complex dependence which it often represents. The talk will detail some approaches for synthetic data generation, and a statistical method for assessing their quality.

# Why Dependent Types Matter?

Thorsten Altenkirch

University of Nottingham, UK
https://scholar.google.com/citations?user=EHksJkUAAAAJ&hl=en

**Abstract.** A dependent type is a type which depends on values. Dependent types are used powerful programming languages which can express any property of a program and they are also used in interactive proof systems like Coq or Lean. I will use the agda system to illustrate the potential of dependent types. I will also highlight some issues which stop dependent types to fulfil their potential.

# Machine Learning with Guarantees

Aleksandar Bojchevski

University of Cologne, Germany
https://scholar.google.com/citations?hl=en&user=F1APiN4AAAAJ

**Abstract.** From healthcare to natural disaster prediction, high-stakes applications increasingly rely on machine learning models. Yet, most models are unreliable. They can be vulnerable to manipulation and unpredictable on inputs that slightly deviate from their training data. To make them trustworthy, we need provable guarantees. In this talk, we will explore two kinds of guarantees: robustness certificates and conformal prediction. First, we will derive certificates that guarantee stability under worst-case adversarial perturbations, focusing on the model-agnostic randomized smoothing technique. Next, we will discuss conformal prediction to equip models with prediction sets that cover the true label with high probability. The prediction set size reflects the models uncertainty. To conclude, we will provide an overview of guarantees for other trustworthiness aspects such as privacy and fairness.

# Why is AI a Social Problem in 2023?

Marija Slavkovikj

University of Bergen, Norway
https://scholar.google.com/citations?hl=en&user=g8UBNwUAAAAJ

**Abstract.** Artificial Intelligence has been an active research area since 1956. In the same timespan AI as an area of innovation and technology has been in and out of existence. The tools we use have always played a role in shaping society, but AI has not so far been discussed as a politically relevant topic. Some of the public discourse today considers topics of super intelligence and machine supremacy. The talk will discuss the reality of AI, what has changed in 2023, the tools that we have available today and the need to decide what kind of socio-technical society do we want to live in.

# Contents

# AI and Natural Language Proccessing

# Extracting Entities and Relations in Analyst Stock Ratings News

Ivan Krstev[✉], Igor Mishkovski, Miroslav Mirchev, Blagica Golubova, and Sasho Gramatikov

Faculty of Computer Science and Engineering, Rugjer Boshkovikj 16, 1000 Skopje, North Macedonia
ivan.krstev.1@students.finki.ukim.mk
https://www.finki.ukim.mk/en

**Abstract.** Massive volumes of finance-related data are created on the Internet daily, whether on question-answering forums, news articles, or stocks analysis sites. This data can be critical in the decision-making process for targeting investments in the stock market. Our research paper aims to extract information from such sources in order to utilize the volumes of data, which is impossible to process manually. In particular, analysts' ratings on the stocks of well-known companies are considered data of interest. Two subdomains of Information Extraction will be performed on the analysts' ratings, Named Entity Recognition and Relation Extraction. The former is a technique for extracting entities from a raw text, giving us insights into phrases that have a special meaning in the domain of interest. However, apart from the actual positions and labels of those phrases, it lacks the ability to explain the mutual relations between them, bringing up the necessity of the latter model, which explains the semantic relationships between entities and enriches the amount of information we can extract when stacked on top of the Named Entity Recognition model. This study is based on the employment of different models for word embedding and different Deep Learning classification architectures for extracting the entities and predicting relations between them. Furthermore, the multilingual abilities of a joint pipeline are being explored by combining English and German corpora. For both subtasks, we record state-of-the-art performances of 97.69% $F_1$ score for named entity recognition and 89.70% $F_1$ score for relation extraction.

**Keywords:** Analysts' Ratings · Financial Data · Information Extraction · Named Entity Recognition · Relation Extraction · Multilingual

## 1 Introduction

Stock markets are one of the leading concepts in today's open economy. They can be defined as a collection of exchanges and trades where shares of companies can be bought, sold or issued[1]. These operations are governed by a set of tight

---

[1] Stock Market, www.investopedia.com/terms/s/stockmarket.asp, Accessed: 2023-07-15.

M. Mihova and M. Jovanov (Eds.): ICT Innovations 2023, CCIS 1991, pp. 3–18, 2024.
https://doi.org/10.1007/978-3-031-54321-0_1

rules and regulations, which opens the opportunity for their analysis by experts in the field, like analysts who aim to predict the price target of the shares of a particular company in the near future based on its financial activities. On the other side, we have investors who invest money in a certain business entity in hopes of making a profit under acceptable risk levels. Investors rely heavily on what experts (analysts) have to say and predict about their company or asset of interest. Investments are typically organized in stock portfolios which balance between expected returns and possible risks. There is an abundant scientific literature regarding stock price prediction, portfolio management, risk assessment, algorithmic trading, etc. Numerous works have explored applications of machine learning for these financial applications, and recently particularly deep learning [4,20]. However, many experts are cautious when applying machine learning algorithms as it has shown mixed performance [4,5].

According to [5], when investing in equities investors can base their decisions either on analyst ratings given by human experts or quantitative ratings generated by machine learning. A question arises of whether investors should trust human wisdom more than the advice of machines. Their quantitative ratings are generated using the random forests algorithm, and they employ the human-generated ratings information by analyzing their sentiment. The results reveal that the analysts ratings outperform the quantitative rating, implying that analysts ratings are much more useful for making good decisions. Another study [24], explores a variety of ways for identifying a feature set for stock classification and emphasizes the significance of analyst ratings for bringing valuable human knowledge of the current stock market situation. Analysts provide their knowledge of trading activity statistics derived from historical data and external factors impacting companies' operations. This information is often biased as analysts are pressured to make more optimistic projections due to relations with investment banks [24], but taking into account the number of analysts per stock and the assumption that not all are related to the same banks, we can assume that the ratings variance cancels a significant amount of the bias. Finally, the authors combine features from a technical and fundamental analysis and pose a classification problem where each stock is labeled as buy, hold, or sell.

Keeping in mind the importance of the analyst ratings, we proceed with the process of extracting information out of them, in a form which can be then utilized more easily. Analysts typically share their publicly distributed expertise in a form of a raw unannotated text, containing key information about a company's shares, price targets, and conclusions, usually in a buy-sell-hold form, implying their suggestions on those particular stocks. However, there are dozens of analysts that analyze one company and there are dozens of companies analyzed by one analyst. Hence, it is useful to have a tool that automatically extracts all the information needed from the raw analysts' ratings into an annotated form without manual effort. Therefore, in our study, we utilize information extraction techniques and map the knowledge in the analysts' ratings in order to build a system that can facilitate analyses of companies' performances, improve predictions of stock prices trends and enhance portfolio management.

The problem of information extraction (IE) from analysts' ratings can be divided into two subtasks, Named Entity Recognition (NER) and Relation Extraction (RE). A recent survey of the state-of-the-art methods for NER and RE can be found in [19], while another survey focusing on deep learning methods for named entity recognition can be found in [15]. NER has a long and illustrious history as a tool for financial texts analysis. Its function is to process text in order to identify specific expressions as belonging to a label of interest [18]. For example, the study in [8] identifies entities such as "invoice sender name", "invoice number" and "invoice date" in business documents, such as invoices, business forms, or emails. In terms of analysts' ratings, there are a few key concepts i.e., entities that need to be retrieved in order to gain information from the raw text. One might be interested in extracting entities including the name of the analyst, the company of interest, and the predicted price target and position as shown in Fig. 1 a). In the figure, NER retrieves the information that there is an analyst "Susan Roth Katzke", a company "Bank of America" and a price target "$ 47.00", and points to their exact location in the text. Nonetheless, we have no way of knowing if "Susan Roth Katzke" is analyzing "Bank of America" or whether the price objective is for the same company. Although in that particular example, there is only one analyst that evaluates one company with a sole price target and their mutual relationship can be taken for granted, in the wilderness of analyst ratings websites, things can get way more complicated and one text can contain information about multiple analysts evaluating multiple companies. Therefore, in order to find the semantic relationship between the entities [30] we need to employ Relation Extraction (RE). RE typically operates on top of NER or any other sequence labeling architecture, although it can be also solved jointly with NER [25,27,28]. However, we choose the first option, as demonstrated in Fig. 1 b), and after NER we proceed with annotating relationships between the entities to obtain the semantics of the raw text. Now, it is clear that not only this rating is written by some analyst "Susan Roth Katzke" and there is a company "Bank of America", but it can be also stated that the analyst is analyzing that particular company and she assigns the price target of "$ 47.00".

There are numerous other applications of automatic information extraction from financial texts using NER [3,7,16,26], or NER & RE [11,31]. In [22], the reader can find an overview of NER and RE applications in financial texts as well as knowledge graphs construction and analysis. Recent papers have also addressed some other related useful information extraction problems. In [12], the authors have assembled and annotated a corpus of economic and financial news in English language and used it in the context of event extraction, while another study in [29], focused on event extraction from Chinese financial news using automated labeling. Another work in [16], solves a joint problem of opinion extraction and NER using a dataset of financial reviews.

To our knowledge, there are not many recent works in NER and RE using analysts ratings data, and the closest research was presented in [11], where the authors collect and annotate a French corpus with financial data that is not required to be analyst ratings and use it to train models for extracting entities

**Fig. 1.** a) Named Entity Recognition annotation for one Analyst's Rating; b) Relation Extraction annotation on top of NER entities, explaining mutual semantic relationships between them.

and their mutual relationships. This study puts an accent on the data collection and preparation for NER and RE. Namely, they collect and manually annotate 130 financial news articles and only perform proof of concept experiments for entities and relations extraction. They base their models on SpaCy v2 [9] and obtained 73.55% $F_1$-score for NER and 55% $F_1$-score for relation extraction. We extend this study by collecting and annotating a multilingual corpus with Analyst Ratings in both English and German and exhaustively utilize them for training the proposed models. Furthermore, we switch to SpaCy v3 [10] and employ a newly developed RE component which eliminates the usage of dependency parser for extracting relations, and thus overcomes the gap between precision and recall noted in [11] and obtain a better overall $F_1$ score. Another research work in [31], addressed the problems of NER and RE jointly using BiGRU with attention in a corpus of manually annotated 3000 financial news articles. However, the authors do not provide many information about the dataset and do not employ a transformer architecture.

The rest of the paper is organized as follows. Section 2 describes the gathering of the dataset used for training. We move on to technical aspects and methodologies for NER and RE and the process of building multilingual language models in German and English in Sect. 3, and in Sect. 4, we give the outcomes and the results. Finally, we summarize this research in Sect. 5.

## 2   Dataset

The data used in this work was obtained using the API provided by City-FALCON[2] from which we pulled approximately 180000 general financial-related texts. Due to the manually intensive work, we labeled only a few more than 1000 of them. The scraped data also contained a considerable amount of texts not related to analyst ratings, and therefore, we employed a keyword filtering strategy to purify the texts. The strategy consisted in manually inspecting common words and phrases occurring in the ratings and discarding all the texts that

---

[2] CITYFALCON, www.cityfalcon.com.

lacked those words. The annotation process was done using the online annotation tool UBIAI[3] which offers intuitive UI for both NER and RE. The whole process was catalyzed by using pre-annotation strategies including pre-annotation dictionaries and trained models.

The longest rating contained 1088 tokens, whilst the shortest had only 9. On average, the ratings were 79 tokens long and 50% of them were longer than 51. The mean distance between two entities that are part of the same relation was 13 tokens, with the longest distance in the dataset being 260.

Table 1 sums up the statistics for the obtained corpus used for training our proposed NER and RE models. All entities, apart from *POSITION* and *ACTION*, have descriptive names. *POSITION* refers to the rating that an analyst gives to certain stocks which might be used as an indicator either to buy, sell or hold the given stock. There are 4 ratings indicating that the analyst believes that the shares should be bought i.e., "Analyst Buy Rating", "Analyst Strong Buy Rating", "Analyst Outperform Rating" and "Analyst Market Perform Rating". The "Analyst Hold Rating" indicates that the analyst believes the shares should be held. On the other hand, there is "Analyst Neutral Rating" where "Neutral" does not refer to a hold position, but rather a position where the analyst hesitates to share any kind of an opinion. Furthermore, there are 3 ratings indicating that the analyst believes the shares should be sold: "Analyst Sell Rating", "Analyst Strong Sell Rating" and "Analyst Underperform Rating". It is important to note that an adjective before the rating describes its intensity, e.g., "Strong Buy" indicates that the analyst is extremely sure that buying the stocks is a good idea. Other descriptive adjectives of this kind can also be found in the analyst ratings and they all equally apply for sell and hold positions.

In the financial world, it is common that analysts change their mind regarding a given position on a rating after conducting more thorough research or obtaining new information related to the company activities. To denote these changes in ratings from the analysts, the *ACTION* entity is used. In our study, we use 4 actions as shown in Table 2, although they can appear with different synonyms in the obtained ratings.

## 3   Methodologies

Sequence labeling problems today are generally approached by using pre-trained Language Models (LMs) as their backbone, whether transformer architectures like *BERT* [6] and *RoBERTa* [17], or other contextual embedding models based on RNNs such as *BiLSTMs* [2]. All of the pre-trained LMs are trained on huge corpora, which makes them as suitable for the financial domain as they are for any other, and allow us to transfer the general knowledge obtained by processing massive amounts of data to the problem at stake, in a procedure known as transfer learning [21]. The output of the pre-trained models is used as an input of an often simpler classification model for determining the final label for each

---

[3] UBIAI, https://ubiai.tools.

**Table 1.** Description and distribution of the entities and relations of the dataset. All of the entities have a descriptive name except POSITION and ACTION. Position refers to the buy-sell-hold concept, but it can usually be found in many other forms like "outperform", "market perform", "strong buy" etc. Action refers to the change the analyst has made in the position i.e. if we go from positive to negative POSITION ("buy" to "hold") we should expect words like "downgrade" or "cut" to be the ACTION. The "*" sign in the relations refers that the entity has priority over the other one, and if both of them are present, only the prioritized one is taken into consideration.

| NAMED ENTITY RECOGNITION | | |
|---|---|---|
| Entity | Description | Num. Instances |
| ANALYST_NAME | Name of a person who analyzes stocks | 1773 |
| ANALYST_COMPANY | Company employing the analyst | 3101 |
| COMPANY | A company that is analyzed | 6739 |
| TICKER | Unique identifier on the stock market for each company | 1155 |
| PRICE_TARGET | Projected future price of the stocks | 2324 |
| PRICE_START | Starting price of the stock before the rating is announced | 941 |
| POSITION | Analyst suggestion whether stocks should be bought, sold or held | 1669 |
| ACTION | Explains the change in position the analyst has made from past analysis | 1039 |

| RELATION EXTRACTION | | | |
|---|---|---|---|
| Relation | From | To | Num. Instances |
| ANALYST_WORKS_AT | ANALYST_NAME | ANALYST_COMPANY | 715 |
| ANALYZES | ANALYST_NAME*, ANALYST_COMPANY | COMPANY | 1612 |
| HAS_TICKER | COMPANY | TICKER | 1073 |
| AFFECTS | PRICE_TARGET, PRICE_START, ACTION, POSITION | COMPANY | 3979 |
| ASSIGNS | ANALYST_NAME*, ANALYST_COMPANY | PRICE_TARGET, PRICE_START, ACTION, POSITION | 3888 |

**Table 2.** The entity ACTION refers to the changes made in the POSITION between two analyses. The aggregation of the ACTION entity to upgrade-downgrade-reiterate-initiate and the POSITION entity to buy-sell-hold is made for the sole purposes of explanation. Note that, in the ratings texts, these entities can occur with different synonyms, forms, and additional adjectives for intensity.

| Action | Change in Position |
|---|---|
| Upgrade | Hold→ Buy, Sell→ Hold, Sell → Buy |
| Downgrade | Buy → Hold, Hold → Sell, Buy → Sell |
| Reiterate | No Change |
| Initiation | Initialize Position |

token. The same concept is also employed in the RE task, such that instead of classifying entities, we are classifying potential relationships between them.

We can think of the embedding model (EM) as the "central dogma" for NLP, where each token of the text is converted into a pertinent product for the machine i.e. a vector of real numbers. When building a joint model for NER and RE, there are two approaches that might be taken into consideration depending on the exact implementation and position of that EM. Namely, the NER and RE components of the joint model can be trained either as a single LM or they can be divided into two stacked LMs. The former approach brings two options. In the first option, the embedding layer is shared by the two components and is updated in a mutual fashion, which leads to multi-task learning and faster training. However, in this way, we have to use the shared embedding for both NER and RE in the future, despite the fact that one model may be superior for NER and another for RE. The second option is to train the two components together, but with separate embedding layers, which will make the training process slower, but on the other hand, it cancels the problem with the performance compatibility of the LMs for NER and RE.

The second approach is to train the two components separately and only connect them during the inference stage. As a result, we can conduct more granular and targeted experiments for NER and RE, while using less GPU resources. However, the time complexity increases, but we gain on the simplicity of the overall architecture. All of the following proposed models in this research follow this particular approach. Furthermore, we follow the work presented in [23], where they discuss two alternatives for NER. The first one is to fine-tune the transformer layer on the NER task and only use a simple linear layer for the token classification, and the second is to directly use the embedding from the pre-trained LMs and employ a more complex classification layer. They arrive to the conclusion that the fine-tuning strategy beats the latter, so we follow their lead, but instead of utilizing only a single linear layer as the model's head, we also utilize very simple classifiers including RNN cells.

### 3.1 Named Entity Recognition

The classification layer used to categorize the tokens into one of the entities is simple, which is an advantage of employing and fine-tuning sophisticated embedding models. The dataset is randomly split into two parts, 90% for training and 10% for testing, and those remain static throughout the fine-tuning process of all models in order to provide a fair comparison. The data was provided in an $IOB$ format (Inside-Outside-Beginning), and it was converted to a $DocBin$ file when training with SpaCy. $IOB$ is considered the standard data representation for NER due to the fact that it introduces 3 different types of tags to denote whether a token is at the beginning, inside, or outside the entity. Let us consider the ANALYST_NAME 3 token entity "Susan Roth Katzke". The token "Susan" will be labeled as $B - ANALYST\_NAME$, denoting the beginning of the entity, and the next two tokens, "Roth" and "Katzke" will get the $I - ANALYST\_NAME$ which stands for inside the entity. If a token is not part of any named entity, then it is marked as $O$ which stands for outside any entity.

For the NER task, we compare two powerful NLP frameworks, SpaCy [10] and FLAIR [1]. We make use of SpaCy's CLI (Command Line Interface) which offers commands for initializing and training pipelines. The pipeline used to perform a NER task consists of an embedding model (transformer) and a NER classifier that consists of a single linear layer as an LM Head to the transformer. Most of SpaCy's proposed hyperparameters were adopted without changes because they have been demonstrated to be quite successful. Additionally, Adam [13] with $warmup$ steps was used as an optimizer with an initial learning rate of $5e - 5$, which allows the tuning of more sensitive parts in the model like the attention mechanism. In our work, we used the case-sensitive variants of five different transformer architectures as an embedding component in the SpaCy pipeline: $bert-base-cased$, $distilbert-base-cased$, $roberta-base$, $xlm-roberta-base$ and $albert-base-v2$.

We used the same pipeline architecture in the FLAIR framework, with the only difference being the additional RNN layer between the embedding model and the linear decoder. According to prior FLAIR research and experiments, adding a single LSTM layer to an IE task like NER has proven to be quite effective. Furthermore, we utilize stacked embeddings [2] which allows us to use a few embedding models at the same time concatenated as a single vector. The initial learning rate for the models was set to 0.1 with an annealing factor of 0.5 on every two epochs without improvement. These values were taken based on similar prior experiments presented in [14] in order to avoid the expensive cost of hyperparameter optimization. We used FLAIR to explore static embedding types such as GloVe and stacked embeddings combining FLAIR forward and backward LMs with either GloVe or BERT embeddings.

## 3.2   Relation Extraction

The embedding component of the RE model is identical to that of the NER model. In the case of NER, after embedding the tokens, each obtained vector is fed as an input to a classification layer. However, because one relation is represented by two entities, and entities can have numerous tokens, a few more steps are required for extracting relations. The first step of relation extraction is to create a matrix $E$ that contains the vectors of each entity in a document:

$$E = \begin{bmatrix} e_{111} & e_{112} & e_{113} & \dots & e_{11n} \\ e_{121} & e_{122} & e_{123} & \dots & e_{12n} \\ e_{211} & e_{212} & e_{213} & \dots & e_{21n} \\ \dots & \dots & \dots & \dots & \dots \\ e_{m11} & e_{m12} & e_{m13} & \dots & e_{m1n} \end{bmatrix},$$

s.t. $n$ is the dimension of the embedding space, and $m$ is the number of entities in the document. Each entity can contain multiple tokens, so in $e_{121}$, the first 1 denotes entity 1 in the document, 2 denotes the second token in the entity, and the second 1 is the index of a single value from that embedding. After defining matrix $E$, we deal with the multi-token named entities and use the average pooling operator in order to obtain a single vector per entity:

$$E' = \begin{bmatrix} E_{11} & E_{12} & E_{13} & \dots & E_{1n} \\ E_{21} & E_{22} & E_{23} & \dots & E_{2n} \\ E_{31} & E_{32} & E_{33} & \dots & E_{3n} \\ \dots & \dots & \dots & \dots & \dots \\ E_{m1} & E_{m2} & E_{m3} & \dots & E_{mn} \end{bmatrix}.$$

After obtaining the matrix $E'$, pairs of entities are mutually combined, representing a potential relation:

$$R = \begin{bmatrix} E_{11} & \cdots & E_{1n} & E_{21} & \cdots & E_{2n} \\ E_{21} & \cdots & E_{2n} & E_{11} & \cdots & E_{1n} \\ E_{11} & \cdots & E_{1n} & E_{31} & \cdots & E_{3n} \\ \cdots & \cdots & \cdots & \cdots & \cdots & \cdots \\ E_{(m-1)1} & \cdots & E_{(m-1)n} & E_{m1} & \cdots & E_{mn} \end{bmatrix}.$$

Each of the rows in matrix $R$ is a vector representation for a possible relation in a document, and as such, it is fed as an input of a LM. The output of that model is a numerical value denoting the probability of the given entity pair (relation) belonging to one of the relation classes: $ANALYST\_WORKS\_AT$, $ANALYZES$, $HAS\_TICKER$, $AFFECTS$ and $ASSIGNS$. Both, SpaCy and FLAIR follow this idea with minor differences and implement it in different Deep Learning libraries, i.e., Thinc and PyTorch respectively. Again, we utilize the transformer architectures with SpaCy, and FLAIR embeddings and GloVe with the FLAIR framework.

Most analyst ratings are written according to some unofficial criteria, and they all have a similar structure, regardless of the analyst or the analyst company they come from. Usually, they are written in a very concise way and are not prone to ambiguity. As a consequence, we have a well-defined text, such that all the entities that are in a mutual relation are close to one another. Our proximity analysis has shown that 95% of the entities that share a relation are within a window of 40-token radius. Therefore, we discarded all the relations that are not within the predefined window and trained only on entity pairs that are considered to be close enough. This also goes hand in hand with the fact, that both architectures, transformer and biLSTMs, have reduced accuracy when predicting long text sequences.

### 3.3 Multilingual Models

In NLP, multilingualism refers to the idea of training a single model using data from multiple languages. These models can subsequently be fine-tuned on a multilingual corpus or a monolingual corpus and used for other languages that are similar i.e. from the same language family, as shown in [14], where a multilingual NER model in Macedonian is trained and later tested on Serbian corpus with some fairly promising results. Following the work presented there, we also test the $xlm - roberta - base$ model trained with analyst ratings in English, on a German corpus containing 100 ratings. The model performed better than a random baseline, achieving 44.8% $F_1$ for NER and 32.61% for RE, however, the results were far from what was achieved for the English ratings. Although both languages come from the Germanic family, they have some fairly different grammar and syntax. Let us look at the verb "zurückstufen" for example. It translates to English "downgrade" and denotes the entity "ACTION" in our use case. In German syntax, this verb splits into two parts, such that "stufen" stays in the second position, whereas "zurück" goes last. These parts cannot be annotated together which forms a kind of ambiguity compared to the English corpus of analyst ratings. We further extended this idea by labeling 100 more analyst ratings in German in order to infiltrate them into the training data and define the special rules for data annotation in German.

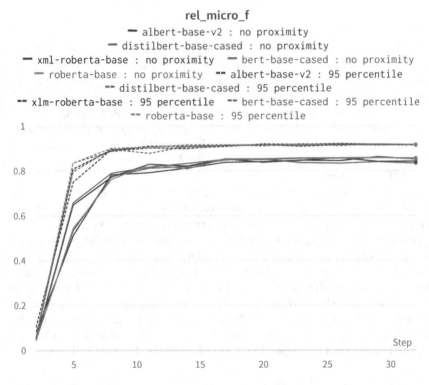

**Fig. 2.** Difference in the $F_1$-score (y-axis) for relation extraction between the transformers trained with proximity (95 percentile) and without proximity.

The multilingual corpus was fed to 3 transformer models, $xlm - roberta - base$, which is the multilingual version of $roberta - base$, pre-trained on 2.5 TB of data in 100 different languages. We also utilized $bert - base - multilingual - cased$ and $distilbert - base - multilingual - cased$ pre-trained on 104 different languages. On the other hand, we also utilize FLAIR forward and backward multilingual embedding models, pre-trained on more than 300 languages.

## 4   Results and Discussion

All proposed models are evaluated with three different metrics: precision, recall, and $F_1$-score, calculated as the harmonic mean of the former two metrics. Just like the training phase, the evaluation is also performed separately for the NER and RE subtasks. However, considering the fact that RE stands on top of NER, the evaluation metrics for RE are obtained using the golden labels from our NER annotations. The test sets make up 10% of the total dataset and include analyst ratings not encountered during the training phase.

## 4.1 Named Entity Recognition

The evaluation ratings were static for each proposed NER model in order to obtain relevant comparisons. In Table 3, evaluation results for the NER task trained with SpaCy are presented.

The results demonstrate that NER with SpaCy achieved state-of-the-art performance with an almost perfect $F_1$-score, which is not surprising given that the entities are not ambiguous, i.e. when analysts talk about downgrading a company, they indeed mean it. The performance of each model individually approves the aforementioned transformer analysis. In this use-case, RoBERTa slightly outperformed BERT ($F_1 = -0.0043$) and XLM RoBERTa ($F_1 = -0.0058$), demonstrating that using a mixed corpus for pre-training did not degrade the model's overall performance for a significant amount. Although DistilBERT achieves the fourth best performance ($F_1 = -0.0072$), this model is on top of the list when it comes to speed and space complexity, and considering real-world applications where system performances matter, it can be considered even as the best candidate. ALBERT achieved the worst results on the analyst ratings dataset. Although having a descent recall, it struggled with precision, especially for the $PRICE\_START$ entity.

**Table 3.** Evaluation of the named entity recognition task using transformer architectures as embedding models with SpaCy.

| Model | NER - SpaCy | | |
|---|---|---|---|
| | Precision | Recall | $F_1$-score |
| roberta-base | **0.9797** | **0.9740** | **0.9769** |
| bert-base-cased | 0.9712 | **0.9740** | 0.9726 |
| xlm-roberta-base | 0.9721 | 0.9702 | 0.9711 |
| distilbert-base-cased | 0.9720 | 0.9673 | 0.9697 |
| albert-base-v2 | 0.8925 | 0.9568 | 0.9235 |

The next set of results for NER come from the FLAIR experiments, presented in Table 4. Even though GloVe is a non-contextual concept for embeddings, based on the co-occurrence matrix of the tokens, it achieves $F_1$-score almost as high as the other transformers and even outperforms ALBERT. Both triplets of stacked embeddings achieve results comparable with $roberta - base$. Surprisingly, the FLAIR embeddings combined with GloVe slightly outperform the combination with BERT. Due to the fact that W&B is not integrated with FLAIR, we omit evaluating the system performances of these models.

## 4.2   Relation Extraction

Relation extraction is a newer and less researched task in the information extraction field and it is yet to acquire the same level of accuracy as NER. Although pre-trained LMs like transformers are also employed as the backbone of RE, it seems that the problem of detecting semantic relationships between entities is more complex than detecting the entities. However, apart from the fact that the RE results are worse than the results obtained for the NER task, we still record high scores for RE, as it can be seen in Table 5.

**Table 4.** Evaluation of the named entity recognition task using static and stacked embedding models with FLAIR ("F" and "B" stand for FLAIR forward and FLAIR backward models).

| Model | NER - FLAIR | | |
|---|---|---|---|
| | Precision | Recall | $F_1$-score |
| GloVe | 0.9754 | 0.9501 | 0.9626 |
| FLAIR F + B + GloVe | **0.9796** | 0.9674 | **0.9734** |
| FLAIR F + B + BERT | 0.9713 | **0.9731** | 0.9722 |

The transformers results presented in Table 5 are based only on the proximity analysis, since it yields 4–6% better $F_1$-scores than the plain models, as seen on Fig. 2. RoBERTa still wins the RE task, however, the relative difference in performance between GloVe for NER (Table 4) and GloVe for RE (Table 5) is inevitable to notice. The reason for this difference is the non-contextual nature of GloVe which manages to extract the entities, but it is not powerful enough to extract the semantic relations between them.

**Table 5.** Evaluation of the RE task with SpaCy (transformers) and FLAIR (GloVe, FLAIR embeddings).

| Model | Relation Extraction | | |
|---|---|---|---|
| | Precision | Recall | $F_1$-score |
| roberta-base | 0.9249 | **0.8707** | **0.8970** |
| bert-base-cased | 0.9249 | 0.8595 | 0.8910 |
| xlm-roberta-base | 0.9230 | 0.8586 | 0.8896 |
| distilbert-base-cased | **0.9325** | 0.8484 | 0.8885 |
| albert-base-v2 | 0.9160 | 0.8521 | 0.8829 |
| GloVe | 0.4673 | 0.3800 | 0.4191 |
| FLAIR F + B | 0.7740 | 0.7726 | 0.7733 |

### 4.3    Multilingual Models

Even though English and German are related, using just multilingual models and training on English corpora does not yield the desired results for German. As a result, infiltrating a portion of the German corpus into the training set with English data is critical. Thus, although the German dataset consisted of only 200 ratings, compared to more than 1000 ratings of the English dataset, when the multilingual models were combined with mixed data, we were able to generate metrics for German that were identical to the English analyst ratings. Table 6 gives an overview of the results for the multilingual models on a German test corpus containing 100 analyst ratings.

**Table 6.** Evaluation of the multilingual named entity recognition models.

| Model | Multilingual NER | | |
|---|---|---|---|
| | Precision | Recall | $F_1$-score |
| bert-base-multi-cased | 0.9564 | 0.9581 | 0.9572 |
| xlm-roberta-base | **0.9701** | **0.9633** | **0.9667** |
| distilbert-base-multi-cased | 0.9532 | 0.9616 | 0.9574 |
| FLAIR Multi F + B | 0.9537 | 0.9354 | 0.9445 |

In Table 7, the results from the multilingual RE models are presented. It is noticeable that the difference between the monolingual and multilingual RE is greater than the monolingual and multilingual NER. We can conclude that more German data are needed in order to obtain identical results for more complex problems like RE.

**Table 7.** Evaluation of the multilingual relation extraction models.

| Model | Multilingual RE | | |
|---|---|---|---|
| | Precision | Recall | $F_1$-score |
| bert-base-multi-cased | 0.7487 | **0.9051** | **0.8195** |
| xlm-roberta-base | 0.7316 | 0.8797 | 0.7989 |
| distilbert-base-multi-cased | 0.7791 | 0.8038 | 0.7913 |
| FLAIR Multi F + B | **0.8471** | 0.7776 | 0.8109 |

We can notice that instead of RoBERTa, the highest $F_1$-score comes from BERT, followed by the multilingual FLAIR embeddings. It is also important to mention that FLAIR records much higher precision than the transformer models and even better scores than the monolingual task.

# 5   Conclusion

In this research paper we go through various points of the employment of analysts ratings in stocks analysis. After perceiving their importance, we proceed toward building an information extraction pipeline consisting of extracting entities (NER) and extracting relations (RE). For that point, more than 1000 analysts ratings in English and 200 ratings in German were manually annotated, forming the first such annotated dataset to the best of our knowledge.

We compare two different NLP frameworks, SpaCy and FLAIR, and explore a few different word embedding possibilities, including transformers, Bi-LSTM-based embeddings, and GloVe in order to maximize the results for both subtasks. Our proposed models obtained state-of-the-art results both for NER (97.69% $F_1$) and RE (89.70%). Furthermore, we explore the system performances of the models and offer a pipeline with an inference time, fast enough for production.

We rounded up this study by examining multilingual models and combing English and German corpora with analyst ratings. Although we had access to only 100 German ratings for training and 100 for testing, the scores for the German ratings were brought extremely close to the ones for English with only 1.02% difference in $F_1$-score for NER and 7.75% for RE.

**Acknowledgement.** We are thankful to CityFALCON for providing us the data and for their collaboration and support. This work was partially funded by the Faculty of Computer Science and Engineering at the Ss. Cyril and Methodius University in Skopje.

# References

1. Akbik, A., Bergmann, T., Blythe, D., Rasul, K., Schweter, S., Vollgraf, R.: Flair: an easy-to-use framework for state-of-the-art NLP. In: NAACL 2019, 2019 Annual Conference of the North American Chapter of the Association for Computational Linguistics (Demonstrations), pp. 54–59 (2019)
2. Akbik, A., Blythe, D., Vollgraf, R.: Contextual string embeddings for sequence labeling. In: COLING 2018, 27th International Conference on Computational Linguistics, pp. 1638–1649 (2018)
3. Alvarado, J.C.S., Verspoor, K., Baldwin, T.: Domain adaption of named entity recognition to support credit risk assessment. In: Proceedings of the Australasian Language Technology Association Workshop 2015, pp. 84–90 (2015)
4. Bartram, S.M., Branke, J., De Rossi, G., Motahari, M.: Machine learning for active portfolio management. J. Finan. Data Sci. **3**(3), 9–30 (2021)
5. Cheng, S., Lu, R., Zhang, X.: What should investors care about? mutual fund ratings by analysts vs. machine learning technique. Machine Learning Technique (August 14, 2021). ADB-IGF Special Working Paper Series "Fintech to Enable Development, Investment, Financial Inclusion, and Sustainability (2021)
6. Devlin, J., Chang, M.W., Lee, K., Toutanova, K.: Bert: pre-training of deep bidirectional transformers for language understanding. arXiv:1810.04805 (2018)
7. Farmakiotou, D., Karkaletsis, V., Koutsias, J., Sigletos, G., Spyropoulos, C.D., Stamatopoulos, P.: Rule-based named entity recognition for greek financial texts.

In: Proceedings of the Workshop on Computational lexicography and Multimedia Dictionaries (COMLEX 2000), pp. 75–78 (2000)

8. Francis, S., Van Landeghem, J., Moens, M.F.: Transfer learning for named entity recognition in financial and biomedical documents. Information **10**(8), 248 (2019)

9. Honnibal, M., Montani, I.: spacy 2: natural language understanding with bloom embeddings, convolutional neural networks and incremental parsing. Unpublished Softw. Appl. **7**(1), 411–420 (2017)

10. Honnibal, M., Montani, I., Van Landeghem, S., Boyd, A.: spaCy: industrial-strength natural language processing in python. Unpublished Softw. Appl. (2020). https://doi.org/10.5281/zenodo.1212303

11. Jabbari, A., Sauvage, O., Zeine, H., Chergui, H.: A French corpus and annotation schema for named entity recognition and relation extraction of financial news. In: Proceedings of The 12th Language Resources and Evaluation Conference, pp. 2293–2299 (2020)

12. Jacobs, G., Hoste, V.: SENTiVENT: enabling supervised information extraction of company-specific events in economic and financial news. Lang. Resour. Eval. **56**, 1–33 (2021)

13. Kingma, D.P., Ba, J.: Adam: a method for stochastic optimization. arXiv preprint arXiv:1412.6980 (2014)

14. Krstev, I., Fisnik, D., Gramatikov, S., Mirchev, M., Mishkovski, I.: Named entity recognition for macedonian language. repository.ukim.mk (2021)

15. Li, J., Sun, A., Han, J., Li, C.: A survey on deep learning for named entity recognition. IEEE Trans. Knowl. Data Eng. **34**(1), 50–70 (2020)

16. Liao, J., Shi, H.: Research on joint extraction model of financial product opinion and entities based on roberta. Electronics **11**(9), 1345 (2022)

17. Liu, Y., et al.: Roberta: a robustly optimized Bert pretraining approach. arXiv:1907.11692 (2019)

18. Mikheev, A., Moens, M., Grover, C.: Named entity recognition without gazetteers. In: Ninth Conference of the European Chapter of the Association for Computational Linguistics, pp. 1–8 (1999)

19. Nasar, Z., Jaffry, S.W., Malik, M.K.: Named entity recognition and relation extraction: state-of-the-art. ACM Comput. Surv. (CSUR) **54**(1), 1–39 (2021)

20. Ozbayoglu, A.M., Gudelek, M.U., Sezer, O.B.: Deep learning for financial applications: a survey. Appl. Soft Comput. **93**, 106384 (2020)

21. Pan, S.J., Yang, Q.: A survey on transfer learning. IEEE Trans. Knowl. Data Eng. **22**(10), 1345–1359 (2009)

22. Repke, T., Krestel, R.: Extraction and representation of financial entities from text. In: Consoli, S., Reforgiato Recupero, D., Saisana, M. (eds.) Data Science for Economics and Finance, pp. 241–263. Springer, Cham (2021). https://doi.org/10.1007/978-3-030-66891-4_11

23. Schweter, S., Akbik, A.: Flert: document-level features for named entity recognition. arXiv:2011.06993 (2020)

24. Singh, J., Khushi, M.: Feature learning for stock price prediction shows a significant role of analyst rating. Appl. Syst. Innov. **4**(1), 17 (2021)

25. Sun, C., et al.: Joint type inference on entities and relations via graph convolutional networks. In: Proceedings of the 57th Annual Meeting of the Association for Computational Linguistics, pp. 1361–1370 (2019)

26. Wang, S., Xu, R., Liu, B., Gui, L., Zhou, Y.: Financial named entity recognition based on conditional random fields and information entropy. In: 2014 International Conference on Machine Learning and Cybernetics, vol. 2, pp. 838–843. IEEE (2014)

27. Wang, Y., Yu, B., Zhang, Y., Liu, T., Zhu, H., Sun, L.: TPLinker: single-stage joint extraction of entities and relations through token pair linking. In: Proceedings of the 28th International Conference on Computational Linguistics, pp. 1572–1582. International Committee on Computational Linguistics, Barcelona, Spain (Online) (2020). https://doi.org/10.18653/v1/2020.coling-main.138, https://aclanthology.org/2020.coling-main.138

28. Wei, Z., Su, J., Wang, Y., Tian, Y., Chang, Y.: A novel cascade binary tagging framework for relational triple extraction. In: Proceedings of the 58th Annual Meeting of the Association for Computational Linguistics, pp. 1476–1488. Association for Computational Linguistics, Online (2020). https://doi.org/10.18653/v1/2020.acl-main.136, https://aclanthology.org/2020.acl-main.136

29. Yang, H., Chen, Y., Liu, K., Xiao, Y., Zhao, J.: Dcfee: a document-level Chinese financial event extraction system based on automatically labeled training data. In: Proceedings of ACL 2018, System Demonstrations, pp. 50–55 (2018)

30. Zhou, G., Su, J., Zhang, J., Zhang, M.: Exploring various knowledge in relation extraction. In: Proceedings of the 43rd Annual Meeting of the Association for Computational Linguistics (ACL 2005), pp. 427–434 (2005)

31. Zhou, Z., Zhang, H.: Research on entity relationship extraction in financial and economic field based on deep learning. In: 2018 IEEE 4th International Conference on Computer and Communications (ICCC), pp. 2430–2435. IEEE (2018)

# MakedonASRDataset - A Dataset for Speech Recognition in the Macedonian Language

Martin Mishev[(✉)], Blagica Penkova, Maja Mitreska, Magdalena Kostoska,
Ana Todorovska, Monika Simjanoska, and Kostadin Mishev

Faculty of Computer Science and Engineering, Ss Cyril and Methodiuos University,
Skopje, North Macedonia
{mishev.martin,blagica.penkova,maja.mitreska.1}@students.finki.ukim.mk,
{magdalena.kostoska,ana.todorovska,kostadin.mishev}@finki.ukim.mk

**Abstract.** Using dataset analysis as a research method is becoming
more popular among many researchers with diverse data collection
and analysis backgrounds. This paper provides the first publicly available dataset consisting of audio segments and appropriate textual transcription in the Macedonian language. It is appropriately preprocessed
and prepared for direct utilization in the automatic speech recognition
pipelines. The dataset was created by students at the Faculty of Computer Science and Engineering as part of the elective course, 'Digital
Libraries', with the audio segments sourced from a YouTube channel.

**Keywords:** dataset analysis · audio segments · transcriptions ·
Macedonian language · speech recognition

## 1 Introduction

Speech recognition is an advanced technology that bridges the gap between spoken language and machine comprehension. It involves the ability of machines to
accurately recognize, transcribe, and interpret human speech into written text
or actionable commands. By analyzing speech signals and employing sophisticated pattern recognition techniques, this interdisciplinary technology empowers
machines to automatically comprehend and process spoken language [6].

The applications of speech recognition are diverse and impressive. One primary use case is speech-to-text transcription, where spoken words are converted
into written text, facilitating efficient note-taking, documentation, and accessibility for individuals with hearing impairments. Furthermore, speech recognition
enables voice commands, allowing users to interact with devices and systems
using their voice, enhancing convenience and hands-free operation. Additionally,

Supported by Faculty of Computer Science and Engineering, Skopje, N. Macedonia.
M. Mishev, B. Penkova, M. Mitreska, M. Kostoska, A, Todorovska, M. Simjanoska, and
K. Mishev —Equal Contribution.

voice search relies on this technology, enabling users to retrieve information or execute tasks by speaking naturally [8].

Speech recognition is a complex and rapidly evolving field that draws upon various disciplines. It combines linguistics to understand the structure and semantics of the human language, computer science to design efficient algorithms and systems, signal processing to analyze and interpret speech signals, and machine learning to train models that can improve accuracy over time. The continual advancements in technology fuel the expansion and enhancement of speech recognition, making it an indispensable tool in numerous industries and applications [4].

Despite being widely used, speech recognition technology faces significant difficulties when it comes to low-resource languages. One of the main challenges for researchers and developers working on speech recognition systems for these languages is the scarcity of linguistic resources and the lack of high-quality data [2].

Macedonian is a prime example of a low-resource language, with limited resources available. The Macedonian language is a Slavic language spoken by approximately two million people, primarily in North Macedonia and neighbouring countries [5]. Regardless of its relatively small population, Macedonian is an important language, both culturally and economically and the demand for speech recognition technology in Macedonian is growing. Still, the lack of easily accessible data presents a significant challenge in developing reliable and accurate speech recognition systems for this language.

A potential solution to tackle this challenge is the creation of a robust dataset specifically tailored for speech recognition in Macedonian. Developing such a dataset can help overcome the obstacles posed by the low-resource nature of the language, enabling better performance and accuracy in speech recognition systems.

This paper presents the MakedonASRDataset[1], a novel dataset specifically developed for speech recognition in the Macedonian language. Our objective is to overcome the limited availability of data for the Macedonian language by offering an extensive collection of audio segments, accompanied by their corresponding transcriptions. The dataset encompasses recordings of various speech types, such as read speech, spontaneous speech, and speech in noisy environments, providing a diverse range of data for training and evaluating speech recognition systems in Macedonian. Throughout this paper, we will describe the dataset creation process and analyze its distinctive characteristics.

The development of the MakedonASRDataset holds significant promise in advancing speech recognition research for the Macedonian language. By providing this dataset, researchers and developers can expedite the creation of highly effective speech recognition systems tailored to Macedonian, thereby enhancing technology accessibility and communication for Macedonian speakers. Additionally, the dataset's availability can foster progress in related fields, such as natural language processing and linguistics, contributing to a deeper comprehension of

---

[1] https://drive.google.com/file/d/1ecgz27gzUTzgwu7Lof6l_cM7PQrN2cAV/view?usp=drive_linkMakedonASRDataset.

the Macedonian language. Furthermore, the dataset offers valuable insights into the phonetics and phonology of Macedonian, facilitating the study of common pronunciation patterns and dialectal variations.

## 2   Related Work

Building high-quality datasets for speech and language technology applications is crucial for the development of robust and accurate models. Several recent works have focused on creating large-scale datasets for various tasks, such as speech recognition, speaker identification, and language modelling. Here are some related works that have focused on creating such datasets: Oscar dataset [3], VoxLingua107 [9], Common Voice [1],TED-LIUM [7], AISHELL

The OSCAR project [3] is an open-source initiative that aims to provide large amounts of unannotated web-based data for machine learning and AI applications. The project has developed efficient data pipelines to classify and filter this data, with a focus on improving its quality and providing resources for low-resource languages, including Macedonian. The goal is to make these new technologies accessible to as many communities as possible. This paper suggested a way to be able to translate using only monolingual data from the source and the target language. The OSCAR dataset, which contains data in 166 different languages, has both original and deduplicated versions of the data available. The Data Splits Sample Size subsection provides information on the size of each sub-corpus, including the number of words, lines, and sizes for both versions of the data, as well as the language code for each sub-corpus.

The main goal of the VoxLingua107 dataset [9] is to explore the potential use of web audio data for identifying spoken languages. The researchers generated search phrases from Wikipedia data specific to each language and used them to retrieve videos from YouTube for 107 different languages. They then employed speech activity detection and speaker diarisation techniques to extract speech segments from the videos. A post-filtering step was carried out to remove segments that were unlikely to be in the given language, increasing the accuracy of labelling to 98% based on crowd-sourced verification. The resulting dataset, called VoxLingua107, contains 6628 h of speech, with an average of 62 h per language, and it comes with an evaluation set of 1609 verified utterances. The researchers used the dataset to train language recognition models for several spoken language identification tasks and found that the results were competitive with using hand-labelled proprietary datasets. The dataset is publicly available for research purposes.

The Common Voice corpus [1] is a large collection of transcribed speech that is designed for use in speech technology research and development, particularly for Automatic Speech Recognition. The project uses crowd-sourcing to collect and validate the data, and the most recent release includes data from 38 languages. Over 50,000 people have participated in the project, resulting in 2,500 h of collected audio, making it the largest audio corpus in the public domain for speech recognition in terms of both hours and number of languages.

The researchers conducted speech recognition experiments using Mozilla's Deep-Speech Speech-to-Text toolkit, applying transfer learning from a source English model to 12 target languages, including German, French, and Italian. The results showed an average improvement in Character Error Rate of $5.99 \pm 5.48$ for these languages, which are the first published results on end-to-end Automatic Speech Recognition for most of them.

The TED-LIUM [7] dataset is widely used corpus for research in automatic speech recognition(ASR). It was created by the authors for their participation in the IWSLT 2011 evaluation campaign, with the goal of decoding and translating speeches from TED conferences. The dataset focuses on English audio and transcripts that have been aligned for ASR tasks. The methodology of this dataset is as follows:

1. Data Collection: The authors collected the data from TED (Technology, Entertainment, Design) conferences, specifically from freely available video talks on the TED website. They developed a specialized tool to extract videos and closed captions from the website.
2. Corpus Size: The TED-LIUM corpus consists of 818 audio files (talks) along with their corresponding closed captions. In total, it provides approximately 216 h of audio data, out of which 192 h are actual speech.
3. Speaker Distribution: The dataset includes contributions from 698 unique speakers. Among these speakers, there are approximately 129 h of male speech and 63 h of female speech.
4. Language: The original language of the TED talks in the dataset is English, making it suitable for English ASR research and related tasks.
5. Alignment: The audio files and the corresponding transcripts are carefully aligned, allowing researchers to use the dataset for tasks such as transcription, ASR system training, and speech recognition evaluations (Fig. 1).

## 3   Methodology

In this section, we will describe the methodology used in our study of speech data in the Macedonian language. Specifically, we will cover the data sources, data cleaning and preprocessing, annotation and labelling tools, quality assurance procedures, and the process of storing the final dataset.

### 3.1   Data Sources

In our study, we collected speech data from YouTube, where we selected videos containing interviews and morning shows in the Macedonian language. The YouTube videos were then converted to MP3 format, and each student was provided with an audio file of approximately 150 min in duration.

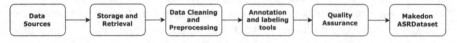

**Fig. 1.** The structure of our proposed methodology

## 3.2   Storage and Retrieval

We used Google Drive to store and share the audio and transcription files. The students were organized into folders, and each student had access to only their own folder. This ensured that the data was kept confidential and that only the students who were authorized to work with a particular dataset had access to it. To request access to a specific folder, students were asked to send an email with their name, surname, and index. This was necessary to ensure that only authorized individuals had access to the data. Once the request was received, it was reviewed to ensure that the folder was suitable for the student. This was done to avoid any potential issues that may arise from unauthorized access or sharing of the data.

## 3.3   Data Cleaning and Preprocessing

While YouTube is a valuable source of speech data, it can be challenging to extract clean and useful audio samples due to extraneous sounds that can interfere with the quality of the speech data. These extraneous sounds include background noise, music, and other sources of interference that can impact the accuracy and reliability of the dataset. Therefore, it is essential to preprocess the data by removing any unwanted noise or artifacts before using it for analysis.

In our study, the students were responsible for cleaning and preprocessing the audio data provided to them. This task involved cutting the audio into smaller segments of 10 to 15 s in length and removing any overlapping conversation or extraneous noise. To achieve the necessary preprocessing of the speech data, the students were provided with tools and guidance to effectively clean and preprocess the audio segments. The students utilized Audacity[2], a free and open-source audio editing software, to cut the longer audio files into shorter, more manageable segments of 10 to 15 s in duration. Additionally, the students were instructed to skip any parts of the audio where multiple speakers were speaking simultaneously, as these segments can be difficult to transcribe accurately. The main goal of this data-cleaning process was to ensure that the resulting dataset was of high quality and suitable for use in subsequent analyses.

## 3.4   Annotation and Labelling Tools

In order to label and annotate the speech data, the students used an Excel spreadsheet that included two columns: segment name and transcription. The transcriptions were required to be written exactly as spoken, regardless of any errors or dialectical differences, and no corrections were allowed to be made. Basic punctuation marks, such as periods, commas, question marks, and exclamation marks, were allowed. In case the speaker mentioned any numerical value, it was transcribed using words.

---

[2] https://www.audacityteam.org/.

### 3.5   Quality Assurance

In our study, quality assurance of the transcribed data was a crucial step to ensure the accuracy and reliability of the dataset. We reviewed each student's transcription and provided a score from 0 to 100, where 100 represented a perfect transcription with no errors or omissions. Additionally, we checked for consistency in the use of spelling and punctuation and made sure that transcriptions accurately reflected the spoken language, even if the speaker used non-standard grammar or dialect. Any errors or inconsistencies were corrected and feedback was provided to the students to improve their transcription skills.

### 3.6   Storing the Final Dataset

After completing the quality assurance process, the final dataset was created by compiling the best-reviewed transcriptions into one large dataset. This dataset will be made available online for anyone who wishes to use it for research or other purposes. By making this dataset public, we hope and intend to contribute to the field of speech research and promote further studies on the Macedonian language.

## 4   Dataset Statistic

Dataset statistics refer to the quantitative information that can be derived from a dataset. This information can include measures of central tendency (such as mean, maximum, minimum and more), measures of spread (such as variance and standard deviation), and other descriptive statistics that can provide insights into the properties and characteristics of the dataset.

The MakedonASRDataset is a significant contribution to the field of speech recognition research, providing researchers and developers with a valuable resource for exploring the Macedonian language. In order to fully understand the dataset and leverage its potential, it is crucial to examine its statistics and gain insights into its characteristics and composition. This section aims to provide a comprehensive overview of the dataset statistics, offering a deeper understanding of its quantity, duration, and other relevant features.

Understanding the dataset statistics is essential for researchers to gauge the scale and scope of the MakedonASRDataset. By analyzing the total number of audio segments, researchers can assess the dataset's volume and its potential for capturing diverse speech patterns and contexts. The statistics related to the total duration of the audio segments allow researchers to understand the temporal distribution of the dataset and identify any potential biases or variations in the speech data. Additionally, the average length of the audio segments provides insights into the typical duration of speech samples within the dataset, which is crucial for designing algorithms and models that can effectively process and analyze speech data.

Furthermore, examining the statistics related to the transcriptions in the MakedonASRDataset offers valuable information about the linguistic content and richness of the dataset. The total number of transcriptions provides an understanding of the dataset's size and the amount of text available for analysis. By assessing the total number of words in all transcriptions, researchers can gauge the lexical diversity and breadth of the dataset. This information is crucial for developing speech recognition systems that can accurately handle a wide range of vocabulary and language variations. Additionally, the average number of words per transcription provides insights into the typical length of transcriptions, guiding researchers in designing models that can effectively process varying transcription lengths.

For the purposes of the paper we only included the average, minimum and maximum statistics for the audio segments and transcriptions (both in terms of words and characters).

All of the statistics were calculated using the Python programming language Python. In the next couple of tables, all of the statistics are represented visually.

In Table 1 are represented the statistics of the audio segments. As we can also see from Table 2 the number of audio segments and the number of transcriptions is the same as expected from the requirements. The average length of the audio segments is 14 s, with a minimum length of 0.4 s and a maximum length of 42 s. These statistics provide information about the quantity, duration, and length distribution of the audio segments in the dataset.

**Table 1.** Total number of audio segments

| Description of the statistic | Result |
|---|---|
| Total number of audio segments | 27672 |
| Total duration of the audio segments(seconds) | 320504 |
| Average length of the audio segments(seconds) | 14 |
| Minimal duration of the audio segment(seconds) | 0.40 |
| Maximum duration of the audio segment(seconds) | 42.91 |

The descriptive statistics for the total number of transcriptions in the MakedonASRDataset are shown in Table 2, which offers important details about the features of the dataset. The dataset's total number of transcriptions, 27,672, is referred to as the "total number of transcriptions". These transcriptions serve as the textual representations of the audio segments in the dataset, capturing the spoken content in written form. The 'Total number of words in all transcriptions' indicates the cumulative number of words found across all transcriptions, which amounts to 874,615 words. Moreover, the number of unique words that appear in the transcriptions is 55,301. This metric highlights the extent of linguistic content captured within the dataset, demonstrating its richness and potential for analysis. The 'Average number of words per transcription' represents the mean number of words present in each transcription, which is calculated as 31.60496. With this we provide insights into the typical length or verbosity of transcriptions, indicating the average amount of spoken content captured within each transcription. Lastly, the 'Total number of chars' refers to the entire number of characters present in all transcriptions, which amounts to 5,030,194 characters. By taking into consideration individual characters rather than simply words, this statistic offers a more comprehensive understanding of the textual content inside the dataset.

**Table 2.** Total number of transcriptions

| Description of the statistic | Result |
| --- | --- |
| Total number of transcriptions | 27672 |
| Total number of words in all transcriptions | 874615 |
| Number of unique words in all transcriptions | 55301 |
| Average number of words per transcription | 31.60496 |
| Total number of chars | 5030194 |

On the other hand, in Fig. 2 we have a pie chart which presents information about the gender distribution of voices in the YouTube audios being discussed. According to the chart, approximately 53% of the voices in the covered YouTube audios are classified as male. This indicates that male voices make up the majority of the audio content analyzed. Around 33% of the voices are classified as female, which suggests that female voices represent a significant portion of the audio content. Additionally, the chart shows that approximately 13% of the voices are classified as noise. This means that there is a portion of the audio content that consists of sounds that are not recognizable or understandable as human voices.

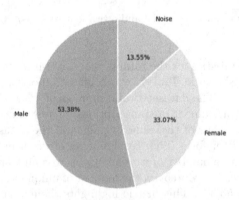

**Fig. 2.** Gender distribution

# 5    Conclusion

In conclusion, this paper has presented the MakedonASRDataset, a valuable resource for speech recognition research in the Macedonian language. The availability of the MakedonASRDataset addresses the scarcity of data for speech recognition research in the Macedonian language, opening up new opportunities for advancements in the field. Since the dataset is publicly available, researchers and developers can leverage it to train and evaluate speech recognition models tailored specifically for Macedonian speakers. This, in turn, promotes technological accessibility and communication improvements for the Macedonian community.

**Acknowledgements.** This work was partially financed by the Faculty of Computer Science and Engineering at the Ss. Cyril and Methodius University in Skopje.

# References

1. Ardila, R., et al.: Common voice: a massively-multilingual speech corpus. arXiv preprint arXiv:1912.06670 (2019)
2. Besacier, L., Barnard, E., Karpov, A., Schultz, T.: Automatic speech recognition for under-resourced languages: a survey. Speech Commun. **56**, 85–100 (2014)
3. Datasets, H.F.: Oscar: a large-scale multilingual benchmark for evaluating cross-lingual transfer (2021). https://paperswithcode.com/dataset/oscar. Accessed: [date]
4. Forsberg, M.: Why is speech recognition difficult. Chalmers University of Technology (2003)
5. Friedman, V.A.: Macedonian: a south slavic language. In: Kortmann, B., van der Auwera, J. (eds.) The Slavic Languages, pp. 441–483. Cambridge University Press (2019)
6. Reddy, D.R.: Speech recognition by machine: a review. Proc. IEEE **64**(4), 501–531 (1976)
7. Rousseau, A., Deléglise, P., Esteve, Y.: Ted-lium: an automatic speech recognition dedicated corpus. In: LREC, pp. 125–129 (2012)
8. Sarikaya, R., Hinton, G.E., Weninger, F.F.: Deep neural network language models for speech recognition. IEEE/ACM Trans. Audio, Speech, Lang. Process. **22**(4), 800–814 (2014)
9. Valk, J., Alumäe, T.: Voxlingua107: a dataset for spoken language recognition. In: 2021 IEEE Spoken Language Technology Workshop (SLT), pp. 652–658. IEEE (2021)

# Bioinformatics

# Explainable Machine Learning Unveils Novel Insights into Breast Cancer Metastases Sites Bio-Markers

Milena Trajanoska$^{(\boxtimes)}$, Viktorija Mijalcheva$^{(\boxtimes)}$, and Monika Simjanoska$^{(\boxtimes)}$

Faculty of Computer Science and Engineering, Ss. Cyril and Methodius University, Rugjer Boshkovikj 16, Skopje 1000, North Macedonia
`{milena.trajanoska,monika.simjanoska}@finki.ukim.mk`,
`mijalchevaviktorija@gmail.com`

**Abstract.** Tumor metastasis is the major cause of cancer fatality. Taking this perspective into account, the examination of gene expressions within malignant cells and the alterations in their transcriptome hold significance in the investigation of the molecular mechanisms and cellular phenomena associated with tumor metastasis. Accurately assessing a patient's cancer condition and predicting their prognosis constitutes the central hurdle in formulating an effective therapeutic schedule for them. In recent years, a variety of machine learning techniques have widely contributed to analyzing empirical gene expression data from actual biological contexts, predicting medical outcomes, and supporting decision-making processes. This paper focuses on extracting important genes linked with each of the most common metastasis sites for breast cancer. Furthermore, the implications of the expression levels of each of the identified sets of bio-markers on the probability of predicting the occurrence of a certain metastasis are illustrated using the Shapley values as a model's explainability framework - an approach that has never been applied on this problem before, unveils novel insights and directions for future research. The pioneering advancements of this research lie in the application of specific feature selection methods and compatible evaluation metrics to produce a small set of bio-markers for targeting a specific metastasis site, and further performing explanatory analysis of the impact of gene expression values on each of the examined metastasis sites.

**Keywords:** Breast Cancer · Explainable Machine Learning · Bio-markers · Gene Expression

## 1 Introduction

Breast cancer is the most prevalent cancer in women, according to the Global Cancer Statistics from 2020 [1], surpassing the number of newly diagnosed cases

---

M. Trajanoska, V. Mijalcheva and M. Simjanoska—These authors contributed equally to this work.

M. Mihova and M. Jovanov (Eds.): ICT Innovations 2023, CCIS 1991, pp. 31–45, 2024.
https://doi.org/10.1007/978-3-031-54321-0_3

from lung cancer. The WHO has recorded approximately 2.3 million cases of breast cancer in 2021 [2]. The observed rates of detection of this type of cancer increase with industrialization and urbanization. Moreover, the rapid development of facilities for early detection are an additional cause of the larger numbers being evidenced each year.

Breast cancer is identified as the state where cells within the breast ducts and lobules become cancerous. The average 5-year survival rate from non-metastatic breast cancer is above 80% [3], whereas the 5-year survival rate for women with metastatic breast cancer is 26% [4]. Approximately half of the women who are diagnosed with breast cancer have no identifiable risk factors, excluding gender and age. Some of the identified factors which are considered to increase the risk of the development of breast cancer include: older age, obesity, family history of breast cancer, harmful use of alcohol, history of radiation exposure, tobacco use, reproductive history, and postmenopausal hormone therapy [2].

A certain investigation postulates that instances of brain metastases manifest in around 10% to 16% of individuals diagnosed with breast cancer. Nevertheless, certain extensive post-mortem studies suggest that the incidences might escalate to levels as notable as 18% to 30%. The majority of these metastatic occurrences are characterized by their rapid emergence, typically transpiring within a span of 2 to 3 years subsequent to the initial diagnosis. Upon the identification of brain involvement, the median survival rate is recorded at 13 months, with less than 2% of patients surviving more than 2 years [5].

Further research is required to determine how many patients with non-metastatic breast cancer later develop any type of metastasis. The drastic decrease of the survival rate in patients with metastatic breast cancer signalizes a significant problem which requires new methodologies and state-of-the-art analyses to discover the factors leading to its occurrence. Moreover, gene expression data has been shown to provide expressive power for predicting tumor status directly. RNA-seq gene expression data is suitable for the analysis of biological samples from various sources, including metastatic cancer cells due to its high sensitivity, wide detection range, and low cost [6].

Currently there are few predictive algorithms for identifying patients who are at risk of developing metastasis from their primary tumor. Furthermore, it is important to identify the factors leading to the cause of metastatic cancer. A lot of effort has been made to discover the relations of genetics and risk factors to cancer state and prognosis prediction. An interesting approach was examined in [7], which compares three data mining algorithms, namely Naïve Bayes, RBF Network, J48 with the goal of predicting breast cancer survivability in two cases: benign and malignant cancer patients, on a large dataset containing 683 breast cancer cases. Based on the results, the Naïve Bayes was the best model achieving 97.36% accuracy on the holdout sample, followed by the RBF Network with 96.77% accuracy, and finally J48 with 93.41% accuracy.

Further research [8] describes a unique methodology involving the utilization of tissue samples obtained through surgical resection from metastatic lung lesions. These samples are compared with gene expression profiles extracted from extra-pulmonary locales within breast cancer patients. The study demonstrates the feasibility of prognosticating lung metastasis in cases of breast cancer by discerning distinct gene expression profiles inherent to organ-specific metastatic lesions.

Additional effort has been placed on developing shared databases of human gene data. One such study [9] presents the HumanNet v2 [10], a database of human gene networks, which was enhanced through the integration of novel data modalities, expansion of data sources, and refinement of network inference algorithms.

The main goal of this study is to identify relevant genes contributing to the metastatic outcome of breast cancer, by using a variety of feature selection techniques. The focus is on implementing Boruta search [11] as a feature selection method to identify a small subset of bio-markers with predictive power for the specific metastasis site. The identified sets of relevant genes for each metastasis site are further proven to have satisfactory expressive power in the task of predicting the target metastasis site. Finally, the extracted relevant bio-marker sets are mapped to their definitions, and an explainable analysis of the impact that they have on the prediction result is provided by using Shapley values as a model's explainability framework [12].

The contributions of our work are as follows:

1. End-to-end Machine Learning (ML) pipeline is implemented including: feature selection, data augmentation, model selection, model hyper-parameter tuning and leave-one-out cross-validation for classifying the presence of a specific metastasis;
2. Explainable analysis is provided for a subset of bio-markers selected with Boruta search using Shapley values as an explainability framework - an experiment that is never conducted before for the problem at hand, and
3. The probe's identifiers of the selected bio-markers are mapped to human genes and their related function is briefly described.

The paper is organized as follows. The data sources for the experiments along with training prediction models, and the explainability framework are described in the following section Methods. The experiments and the results from the methodology are presented in the section Results and Discussion. The highlights of the research achievements are presented in the final section Conclusion.

## 2 Methods

This section describes the data collection and analysis process, as well as the experimental setup and implemented algorithms.

### 2.1 Data Collection

The data for this study was collected from the Human Cancer Metastasis Database (HCMDB) [13], which is a freely accessible database that consists of cross-platform transcriptome data on metastases.

The data collection process was executed with the use of the metadata tables, which consist of metadata on patients, cancer types, metastasis status, metastasis sites, experiments and their identifiers. From all of the cancer patients, only gene expression data on the subset of patients with breast cancer who had a registered metastasis site, and patients who had non-metastatic breast cancer, was retrieved from the National Center for Biotechnology Information (NCBI) Gene Expression Omnibus (GEO) database [14].

The HCMDB database contained additional patients' gene expression data from The Cancer Genome Atlas (TCGA) [15], which were not included in the gathered data set, since after careful inspection, we concluded that 98% of the patients with breast cancer as a primary tumor, for which gene expression data was available, had no information about the metastasis status nor metastasis site. Because the method of calculating gene expression values differed between the NCBI GEO and TCGA data, we decided to collect only patient gene expression data available in the GEO database.

Another minor obstacle was the fact that the gene expression data available in the GEO database was calculated using different methods for different patients. Out of a total of 3521 patients with breast cancer, a maximum-sized subset of 423 patients using the same gene expression micro-array technology was identified. The largest subset of patient's gene expression data used the Affymetrix microarray technology [16]. Affymetrix microarray technology has been successfully validated in our previous research at which we modeled gene expression data to predict colorectal cancer at different stages [17–19].

A total of 54675 probes were recorded for each patient using this technology. The collected samples of each patient had recorded one or more of the following metastasis sites: bone, brain, breast (different breast, same breast different tumor), lung and other. These are the target variables for optimization in the current study.

After collecting the available data with the desired gene expression probes, a data set was created consisting of 54679 columns and 423 rows.

## 2.2   Data Preprocessing

Some of the collected gene expression data were calculated using RMA normalized signal intensity, but another subset of the gene expression data was generated using MAS 5.0 signal intensity processing. After some research a study was identified showing that the logarithm of RMA signal intensity and the logarithm of MAS 5.0 intensity have a good correlation of 0.9913 [20]. For this reason to obtain as much data as possible, a logarithmic transform of the MAS 5.0 values of gene expression was computed and then it was used together with the normalized RMA values. The logarithm of the MAS 5.0 values was not in the lower range [0, 2) for any gene expression that was gathered, meaning it did not belong to the region that had higher variance between the two methods, according to the previously mentioned study.

Hereafter, the standard scaling method was used to scale the values to have unit variance and a mean equal to zero. This transformation was made to achieve a faster convergence of the learning process of some scale dependent models used in the experiments showed later in the study, including the support vector machine, and neural network algorithms. Scaling was not used for training tree based algorithms because these kind of algorithms are not sensitive to the scale of the feature values.

## 2.3   Boruta Search Feature Selection

The Boruta search implements a novel feature selection algorithm for identifying all of the relevant variables in a data set. The algorithm is designed as a wrapper around tree-based classification algorithms and has never been applied for the problem at hand. It iteratively removes the features which are proved by performing a statistical Z-test to be less relevant than their randomly permuted variants. In Boruta, features do not compete among themselves, instead they compete with a randomized version of them, called a shadow feature [11].

In this study, the Boruta search was performed using 5-fold cross-validation on the entire data set of gene expressions, while independently optimizing each of the 5 classes of metastasized tumors: bone, brain, breast, lung and other, and on tumors without metastasis. In this stage of the pipeline, it is important to note that data augmentation (introduced later in the training process) was not performed with the reason to consider the real samples of patients, and not bias the algorithm with synthetically generated data. For each of the 6 target variables, a separate XGBoost [21] model was trained on four of the five folds and evaluated on the one fold which was left out. This process is repeated until each fold becomes the test set and the results are averaged over all test folds. The method is repeated in 10 iterations. The evaluation metric for feature selection was Shapley values. The Shapley values is a widely used explainability framework which attributes the prediction of an ML model to its base features [22]. The explanation of a prediction is performed by assuming that each feature value of the sample represents a player in a game where the prediction is the outcome [23]. We also use this method to explain the influence of the most important features (bio-markers) for each metastasis site. Eventually, for each of the six metastasis sites (bone, brain, breast, lung, other, and no metastasis), a gene profile was generated from the wrapper method for feature selection.

An example of a probe importance plot for bone metastasis is represented in Fig. 1. The probes colored in green represent the real features with larger importance than the shadow feature with maximum importance. For reference, the shadow features with maximum, minimum, mean and median importance are colored in blue.

Such plots were generated for each metastasis site and the selected bio-markers marked as most relevant for each metastasis site are shown in the Supporting Information Table A1. [24]. Each type of metastasis is associated with a list of identified bio-markers or selected probes. The selected bio-markers in the list are ordered by their descending importance.

As it can be seen from the generated gene profiles, each probe appears in only one profile and each profile is unique, which might indicate that it is a

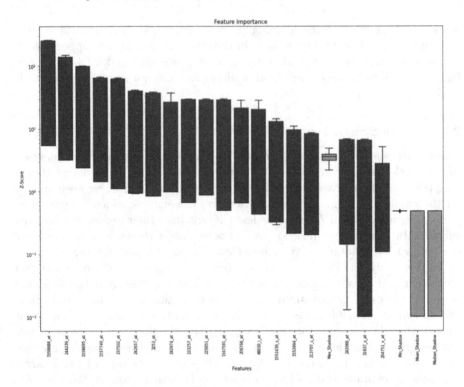

**Fig. 1. Probe importance plot for bone metastasis.** The Y-axis represents the Z-Score for the most relevant probes, and the shadow features which are of maximum, median, mean, and minimum importance. The Z-Scores for the actual probes are represented with box plots in green color, and the Z-Score box plots of the shadow features are represented in blue. Only the probes with importance greater than the importance of the maximum shadow feature are kept using this feature selection method. (Color figure online)

specific gene related to the metastasis of breast cancer to the corresponding site. However, when mapping the probes to their corresponding genes, there is one overlap for the surfactant protein C (SFTPC) in the lung metastasis bio-markers set and the other metastasis bio-markers set. Nevertheless, the probes corresponding to this gene are different, meaning that they are related to different subsequences of this gene.

## 2.4  Experimental Setup

Using the raw gene expression values for the bio-markers obtained from the Boruta search feature selection method, our intention is to test the predictive power of those bio-markers for each metastasis site. One-hot label encoding was performed to convert the metastasis site labels into binary targets expressing the presence/absence of a certain metastasis site, encoded with 1 or 0, respectively.

The values for the metastasis site belong to the set {'brain', 'bone', 'breast', 'lung', 'no', 'other'}. The reason for including 'no' metastasis as a separate class is the fact that while recording patients as having other metastasis, not all possible metastasis sites might have been encountered, thus it is not correct to conclude that if a patient is predicted to not have any of the above-mentioned metastasis sites, then the patient has non-metastatic breast cancer. The data set is multi-target, meaning that one patient might have more metastasis site labels equal to 1.

While performing exploratory data analysis, a severe class imbalance was detected for all of the metastasis types, except for 'other' metastasis and 'no' metastasis. The distribution of the class imbalances per metastasis site can be examined in Table 1.

**Table 1. Class imbalances per metastasis site.** This table represents the distribution of positive samples (meaning that the metastasis is present) and negative samples (meaning that the metastasis is absent) per each patient in the dataset. A severe class imbalance can be spotted for all metastasis sites, except for 'other' metastasis and 'no' metastasis.

| Metastasis type | Num. patients with metastasis | Num. patients without metastasis |
|---|---|---|
| bone | 20 | 403 |
| brain | 46 | 377 |
| breast | 19 | 404 |
| lung | 12 | 411 |
| no | 139 | 284 |
| other | 195 | 228 |

For this reason, it was necessary to balance the class distributions to avoid biasing the models to the majority class. Oversampling and under-sampling techniques were applied on the data set at each cross-validation step consequently considered when training the intelligent classification models. A strategy for mitigating imbalanced data sets involves the over-sampling of the minority class. The most elementary technique involves replicating instances from the minority class, even though these samples do not introduce new value to the model. Instead, a more promising approach is to synthesize new samples by utilizing the existing ones. This approach constitutes a form of data augmentation targeted at the minority class, and is specifically denoted as the Synthetic Minority Oversampling Technique (SMOTE) [25]. SMOTE represents an over-sampling technique in which the minority class is over-sampled by creating synthetic samples, instead of over-sampling with replacement. The algorithm first selects a random sample from the minority class, then it identifies the k nearest neighbors of this sample and chooses one of the nearest neighbors from which it constructs a vector to the selected sample. Then a random point on the vector is chosen to be added to the new over-sampled data set.

An alternative method involves the under-sampling (RandomUnderSampler) technique [26]. This technique entails the random removal of instances from the majority class within the training dataset. Consequently, the quantity of majority class samples is reduced within the modified training dataset, subsequently improving the balance between positive and negative samples. This procedure can be reiterated until the aspired class distribution is attained, potentially resulting in an equilateral number of samples across each class.

The complete Machine Learning pipeline based on the XGBoost method is depicted in Fig. 2. Lead by the intention that false positive classifications are preferred over the false negative for the problem at hand, the hyper-parameter space search is performed using 5-fold cross-validation [27], whilst maximizing the precision of the classifier as the main metric. Thus, a low range of values was chosen for the parameters max_depth and min_child_weight, because the data set does not contain many data points, and over-fitting can easily occur.

The hyper-parameter space search was performed using 5-fold cross-validation for each target metastasis separately, including the 'no' metastasis. We have only used the selected bio-markers from the Bortua search to tune the hyper-parameters, because in the previous step we had identified them as the most significant. In addition, because in total there were more than 54 000

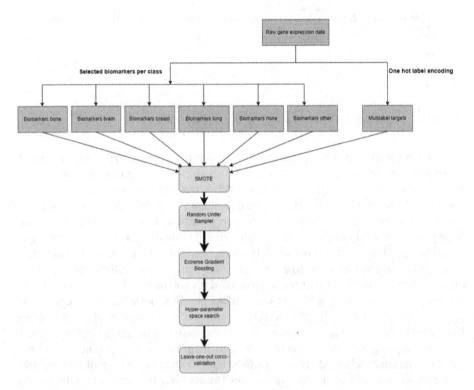

**Fig. 2. Machine Learning pipeline.** The blue rectangles represent data, the orange rounded rectangles represent methods and models, and the arrows represent the flow of the data, for which some actions might be performed, indicated by a label on the arrow. The bolder arrows represent multiple data flows between these steps. (Color figure online)

probes, utilizing all of them for tuning the hyper-parameters would inevitably lead to model overfitting. We have implemented an exhaustive grid search methodology for evaluating the best hyper-parameters while optimizing the precision as the target metric. We have set the options of the hyper-parameters to ranges of low values since the number of patients in the dataset was around 400, which means that addding more depth or estimators might result in fast over-fitting.

After the hyper-parameter space search was performed, model training was executed using leave-one-out [28] cross-validation for each target metastasis, on datasets consisting of the appropriate sets of bio-markers that were selected by the Boruta method. The training was performed on models initialized with the identified hyper-parameters from the previous stage. The leave-one-out cross-validation was chosen with the goal of computing the pessimistic errors of the models, including each patient of the data set. During this process the precision, recall, f1 score, and ROC AUC score are calculated for each of the predicted samples using the leave-one-out method. To ensure that we exclude synthetically generated samples from the leave-one-out process, we augment the data on every fold generated for training the models, instead of first augmenting the data set and then performing the cross-validation. At each step, one sample of the original data set is extracted as a test sample, and all of the other cases are used as a training data set. The training data set is then augmented using the previously mentioned approach with SMOTE. The steps are performed until each case of the original data set has been used as a test sample.

## 3    Results and Discussion

The methodology presented in the previous section, i.e., the Boruta search feature selection together with the generated XGBoost models from the hyper-parameter space search and the leave-one-out cross-validation method produced the results presented in Table 2.

**Table 2. Results from the XGBoost Classifier.** The first column presents the metastasis site, the other columns refer to the evaluation metrics. The lowest scores are outlined in red color.

| Metastasis site | Precision | Recall | F1 Score | ROC AUC |
|---|---|---|---|---|
| bone | 1.0 | 1.0 | 1.0 | 1.0 |
| brain | 0.972 | 0.761 | 0.854 | 0.879 |
| breast | 0.296 | 0.421 | 0.348 | 0.687 |
| lung | 1.0 | 0.917 | 0.956 | 0.958 |
| no | 1.0 | 1.0 | 1.0 | 1.0 |
| other | 0.874 | 0.995 | 0.930 | 0.936 |

Generally, from the table it can be concluded that the selected bio-markers by using the Boruta search method are satisfactory for predicting the metastasis site in most cases, without any notable imbalance between the precision and the recall.

The lowest scores are obtained when predicting breast cancer metastasizing to another location in the same breast, or on the other breast. For this type of metastasis only 3 bio-markers were deemed significant by the implemented feature selection method. These results are outlined with red color in the table. We hypothesize that this is due to the fact that the gene expressions from breast cancer as a primary tumor are not discriminative enough to conclude if there would be a breast metastasis or not.

More importantly, it is significant that only one bio-marker '1552768_at' was identified as important for detecting whether the cancer has metastasized or not. This bio-marker can be tested to identify the risk of metastasis first, and if confirmed positive, the other models can be used to detect the specific site. For 'no metastasis' all of the scores of the model: precision, recall, f1 score, and ROC AUC were calculated to equal 1.0.

### 3.1   Explaining Bio-Markers Impact on Metastasis Prediction

This section presents a comprehensive elaboration of the brain metastasis case by using Shapley values. The other metastasis types investigated in the paper have undergone the same analysis and the results are provided in the Supporting Information section.

Figure 3 presents the importance of each of the selected bio-markers for predicting the outcome of brain metastasis. The probes are ordered according to their importance in the prediction, most important to least important. As it is evident from the figure, the most important probe has the id '1563882_a_at', for which very high gene expression values result in predicting a higher value for the target variable, in this case the existence of brain metastasis.

More accurately, from the dataset used, it can be concluded that high gene expression of the gene corresponding to the probe '1563882_a_at' was found in many patients with brain metastasis. On the other hand, very low and moderate gene expression values of the same probe, result in predicting that brain metastasis has not occurred. In contrast, low gene expression of '218620_s_at' is associated with predicting the occurrence of brain metastasis from breast cancer, and is consistent throughout the entire data set. Likewise, higher expression of the same gene is correlated with predicting that a metastasis to the brain has not occurred.

Other important bio-markers such as '224836_at', '221331_x_at', '1565801_at', '221559_s_at' are also present in high expression on brain metastasis. On the other hand, the bio-markers '2218620_s_at', '222834_at' and '200785_s_at' are directed in the opposite direction. These bio-markers' high expression influences the model to predict the absence of brain metastasis. Most

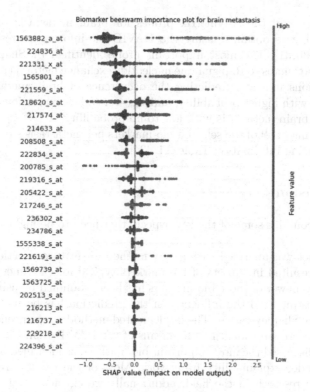

**Fig. 3. Bio-marker beeswarm importance plot for brain metastasis classifier.** Elevated expression levels of probes within the dataset are denoted by red data points, while low expression levels of these same probes are signified by blue data points. The X-axis values indicate the magnitude and orientation of influence that each gene expression value wields over the prognostication of brain metastasis. Positive values, situated to the right of 0, correspond to an increase on the likelihood of predicting the occurrence of brain metastasis in patients. Conversely, negative values positioned to the left of 0 indicate a diminishing effect on the predictive probability of brain metastasis. (Color figure online)

of the other bio-markers shown on the figure contribute to positive prediction of the model and their influence is not that strong as the first ones explained. Figures for each individual model are available in the Supporting Information Fig. 4.

## 3.2  Mapping Metastasis Bio-Markers to Gene Names

To better explain the impact each of the gene expressions has on predicting the brain metastasis site, Affymetrix probes were mapped to gene names. Once again the mapping for the other metastasis types can be found in Supporting Information Appendix A. [29].

The mapping of the Affymetrix probes for brain metastasis presented in Table A1. [24] was obtained using the DAVID bio-informatics resources and online tools [30,31]. The most important probe identified by Shapley values: '1563882_a_at' maps to Rap guanine nucleotide exchange factor 5 (RAPGEF5). High expressions indicate predicting the occurrence of brain metastasis from breast cancer with higher probability, and vice-versa, low expressions contribute to predicting brain metastasis with low to no probability.

The mapping of all of the selected bio-markers per metastasis site is available in the Supporting Information Table A1. [24].

## 4    Conclusion

This section contains some of the key aspects that need to be highlighted in the study.

The dataset was analysed with a new method for feature selection - Boruta search, that resulted in subsets of bio-markers typical for each metastasis site. The predictive power of those bio-markers has been examined by using Extreme Gradient Boosting, and the influence on the prediction has been measured by calculating the Shapley values. The implemented methodology unveiled new bio-markers that were later identified by means of the DAVID tool.

We have discussed the most important bio-markers, as identified by the Shapley values model explainability framework. Most of the results are consistent with previous research in the field, additionally we extract novel bio-markers and associations which have not been characterized before, and we show that these bio-markers have discriminative power for identifying metastasis sites from breast cancer as the primary tumor.

The practical implications of this study include the identification of a small set of discriminative bio-markers for different types of breast cancer metastasis, which can be used for lowering the amount of resources and time needed to analyze gene expression data of patients in order to identify breast cancer metastasis. Moreover, this study extends the current theory by proving the relation between the known bio-markers with specific metastasis sites, and identifying novel bio-markers which have not been studied yet. Further research is needed in order to quantify their significance and deeper relation with the relevant metastasis sites.

Ultimately, this study was limited by the amount of data publicly available for metastasized breast cancer. Another limitation was the different methods of calculating gene expression from cancer tissue, which were used for different patients. The largest subset of patients containing information about metastasis, which we could extract, contained a total of 423 patients. Furthermore, the extreme class imbalance impacted the results of these methods, although synthetic data was generated in order to overcome this obstacle, having more data about real patients with metastatic breast cancer would produce more accurate results and bio-markers.

# Appendix A Supporting Information

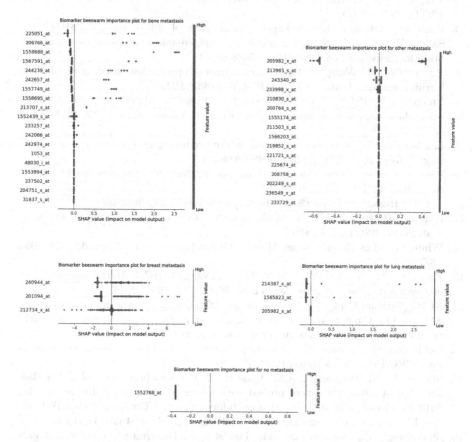

**Fig. 4. Bio-markers importance plot for each target metastasis.** Elevated expression levels of probes within the dataset are denoted by red data points, while low expression levels of these same probes are signified by blue data points. The X-axis values indicate the magnitude and orientation of influence that each gene expression value wields over the prognostication of the target metastasis. Positive values, situated to the right of 0, correspond to an increase on the likelihood of predicting the occurrence of the target metastasis in patients. Conversely, negative values positioned to the left of 0 indicate a diminishing effect on the predictive probability of the target metastasis. (Color figure online)

# References

1. Sung, H., et al.: Global cancer statistics 2020: globocan estimates of incidence and mortality worldwide for 36 cancers in 185 countries. CA: A Cancer J. Clin. **71**(3), 209–249 (2021)

2. WHO: World Health Organization: Breast Cancer (2021). https://www.who.int/news-room/fact-sheets/detail/breast-cancer
3. Sun, Y.-S., et al.: Risk factors and preventions of breast cancer. Int. J. Biol. Sci. **13**(11), 1387 (2017)
4. Peart, O.: Metastatic breast cancer. Radiol. Technol. **88**(5), 519–539 (2017)
5. Salhia, B., et al.: Integrated genomic and epigenomic analysis of breast cancer brain metastasis. PLoS ONE **9**(1), 85448 (2014)
6. Xu, Y., Cui, X., Wang, Y.: Pan-cancer metastasis prediction based on graph deep learning method. Front. Cell Dev. Biol. **9**, 1133 (2021)
7. Chaurasia, V., Pal, S., Tiwari, B.: Prediction of benign and malignant breast cancer using data mining techniques. J. Algorithms Comput. Technol. **12**(2), 119–126 (2018)
8. Landemaine, T., et al.: A six-gene signature predicting breast cancer lung metastasis. Cancer Res. **68**(15), 6092–6099 (2008)
9. Hwang, S., et al.: Humannet v2: human gene networks for disease research. Nucleic Acids Res. **47**(D1), 573–580 (2019)
10. Net, H.: Human Net tool (2021). http://www.inetbio.org/humannet
11. Kursa, M.B., Rudnicki, W.R., et al.: Feature selection with the boruta package. J. Stat. Softw. **36**(11), 1–13 (2010)
12. Winter, E.: The shapley value. Handb. Game Theory Econ. Appl. **3**, 2025–2054 (2002)
13. Zheng, G., Ma, Y., Zou, Y., Yin, A., Li, W., Dong, D.: HCMDB: the human cancer metastasis database. Nucleic Acids Res. **46**(D1), 950–955 (2018)
14. NCBI: National Center for Biotechnology (2021). https://www.ncbi.nlm.nih.gov
15. TCGA: The Cancer Genome Atlas (2021). https://www.cancer.gov/about-nci/organization/ccg/research/structural-genomics/tcga
16. Gohlmann, H., Talloen, W.: Gene Expression Studies Using Affymetrix Microarrays. CRC Press, Boca Raton (2009)
17. Simjanoska, M., Bogdanova, A.M., Popeska, Z.: Bayesian posterior probability classification of colorectal cancer probed with affymetrix microarray technology. In: 2013 36th International Convention on Information and Communication Technology, Electronics and Microelectronics (MIPRO), pp. 959–964 (2013). IEEE
18. Simjanoska, M., Bogdanova, A.M., Popeska, Z.: Recognition of colorectal carcinogenic tissue with gene expression analysis using Bayesian probability. In: Markovski, S., Gusev, M. (eds.) ICT Innovations 2012. AISC, vol. 207, pp. 305–314. Springer, Cham (2012). https://doi.org/10.1007/978-3-642-37169-1_30
19. Simjanoska, M., Bogdanova, A.M., Popeska, Z.: Bayesian multiclass classification of gene expression colorectal cancer stages. In: Trajkovik, V., Anastas, M. (eds.) ICT Innovations, 2013. AISC, vol. 231, pp. 177–186. Springer, Cham (2013). https://doi.org/10.1007/978-3-319-01466-1_17
20. Millenaar, F.F., Okyere, J., May, S.T., Zanten, M., Voesenek, L.A., Peeters, A.J.: How to decide different methods of calculating gene expression from short oligonucleotide array data will give different results. BMC Bioinform. **7**(1), 1–16 (2006)
21. Chen, T., He, T., Benesty, M., Khotilovich, V., Tang, Y., Cho, H., et al.: Xgboost: extreme gradient boosting. R package version 0.4-2 **1**(4), 1–4 (2015)
22. Nowak, A.S., Radzik, T.: The shapley value for n-person games in generalized characteristic function form. Games Econom. Behav. **6**(1), 150–161 (1994)
23. Roth, A.E.: The Shapley value: essays in honor of Lloyd S. Cambridge University Press, Cambridge (1988)

24. Trajanoska, M., Mijalcheva, V., Simjanoska, M.: Affymetrix probes to gene names mapping. https://github.com/MilenaTrajanoska/explainable-ml-breast-cancer-metastases-bio-markers/blob/main/Supporting%20Information/A3.%20Affymetrix_probes_to_gene_names_mapping.pdf

25. Chawla, N.V., Bowyer, K.W., Hall, L.O., Kegelmeyer, W.P.: Smote: synthetic minority over-sampling technique. J. Artif. Intell. Res. **16**, 321–357 (2002)

26. Yen, S.-J., Lee, Y.-S.: Under-sampling approaches for improving prediction of the minority class in an imbalanced dataset. In: Huang, D.S., Li, K., Irwin, G.W. (eds.) Intelligent Control and Automation. LNCIS, vol. 344, pp. 731–740. Springer, Cham (2006). https://doi.org/10.1007/978-3-540-37256-1_89

27. Browne, M.W.: Cross-validation methods. J. Math. Psychol. **44**(1), 108–132 (2000)

28. Webb, G.I., Sammut, C., Perlich, C., et al.: Lazy Learning. Encyclopedia of Machine Learning. springer us (2011)

29. Trajanoska, M., Mijalcheva, V., Simjanoska, M.: Mapping metastasis bio-markers to gene names

30. Huang, D.W., Sherman, B.T., Lempicki, R.A.: Systematic and integrative analysis of large gene lists using David bioinformatics resources. Nat. Protoc. **4**(1), 44–57 (2009)

31. Huang, D.W., Sherman, B.T., Lempicki, R.A.: Bioinformatics enrichment tools: paths toward the comprehensive functional analysis of large gene lists. Nucleic Acids Res. **37**(1), 1–13 (2009)

# Implementation of the Time Series and the Convolutional Vision Transformers for Biological Signal Processing - Blood Pressure Estimation from Photoplethysmogram

Ivan Kuzmanov[1], Nevena Ackovska[1], Fedor Lehocki[2],
and Ana Madevska Bogdanova[1(✉)]

[1] University Ss. Cyril and Methodius, Faculty of Computer Science and Engineering,
Rugjer Boskovikj 16, 1000 Skopje, North Macedonia
`ivan.kuzmanov@students.finki.ukim.mk`,
`{nevena.ackovska,ana.madevska.bogdanova}@finki.ukim.mk`
[2] Institute of Measurement Science, Slovak Academy of Sciences and Faculty of
Informatics and Information Technologies, STU, Bratislava, Slovakia
`fedor.lehocki@stuba.sk`

**Abstract.** Blood pressure estimation is crucial for early detection and prevention of many cardiovascular diseases. This paper explores the potential of the relatively new transformer architecture for accomplishing this task in the domain of biological signal processing. Several preceding studies of blood pressure estimation solely for PPG signals have had success with CNN and LSTM neural networks. In this study two types of transformer variants are considered: the time series and the convolutional vision transformers. The results obtained from our research indicate that this type of approach may be unsuitable for the task. However, further research is needed to make a definitive claim, since only simple transformer type are considered.

**Keywords:** transformers · PPG · blood pressure · time series transformer (TST) · convolutional vision transformer (CvT)

## 1 Introduction

Transformers are a type of neural network architecture. For the first time they were introduced in a paper "Attention is all you need" (Vaswani et al.) published in 2017 [1]. Since then, they have found a wide range of uses in different domains, including the processing of biological signals. The transformer architecture is suitable for tracking both short and long term dependencies in time series data. This architecture offers several improvements over the more traditional ones such as the ability to reuse in the form of transfer learning, greater interpretability, parallel processing, shorter training times, and fewer parameters

M. Mihova and M. Jovanov (Eds.): ICT Innovations 2023, CCIS 1991, pp. 46–58, 2024.
https://doi.org/10.1007/978-3-031-54321-0_4

[1]. Several hybrid model architectures have been developed for specific tasks. In this paper, we will consider two of them: specifically the time series transformer and the convolutional vision transformer. These types of neural network are still relatively new and while promising results have been achieved, the full potential of transformers in biosignal processing is still being explored.

The human body is an elaborate system made up from several intricate systems. These systems interact with one another further increasing the complexity. Biological signals, or biosignals for short, are signals generated by living organisms in response to internal or external stimuli. The biosignals preform vital duties in the various physiological process, such as: sensory perception, communication, and control of bodily functions. These signals don't have to be electrical, they can also be chemical, produced by the hormones or mechanical produced by the muscles [2]. Various techniques have been developed to measure the signals generated by the human body, including: abnormality detection, diagnosis, treatment monitoring, and research purposes.

There are several steps that may be undertaken while preprocessing a signal, such as denoising, feature extraction, feature selection and transformation [3]. While these steps are not exclusive to biosignals, special care must be taken during their analysis to account for patient-specific biosignal structures. Some of the most frequently measured biological signals include: EMG, EEG, ECG and PPG.

ECG, Electrocardiogram, is recording of the heart's electrical activity. It's a commonly measured biosignal that has many uses, some of the most popular are: measuring the heart rate, examining the rhythm of heartbeats, diagnosing heart abnormalities, emotion recognition and biometric identification [3]. PPG, Photoplethysmogram, is a biosignal of the volumetric changes of blood in the body's periphery. It's important to note that PPG is measured using a low-cost sensor and the process is totally non-invasive [4]. These characteristics make this type of signal an interesting feature for various experiments. One example is the process of blood pressure estimation, which is a measurement of critical importance and is usually measured along side the vital signs.

In this paper, we investigate the use of a hybrid transformer as a model architecture for estimating blood pressure from PPG. The paper is structured as follows: in the first section,"Introduction", we delve into the domain of our study and provide a brief literature review on the subjects of interest. In the second section, "Methodology", we clearly define our objectives and evaluation methods, and describe our models and data collection methods. In the third section, "Results", we present the results of our study, including the classification results for the heartbeat classification problem and the regression results for the blood pressure estimation problem. At the end of this section, we discuss the significance of our findings. Finally, the paper concludes with the fourth section providing a summary of our work highlighting our main contributions.

## 1.1  Literature Review

The domain of cuffless blood pressure estimation and hypertension detection from ECG and/or PPG is investigated by several papers, most of which are built on datasets that are yet to be made public. Many factors influence blood pressure, including the patient's age, gender, race, family history, lifestyle habits, medications, and diseases [5]. The prevalence of hypertension makes the estimation of blood pressure a popular research topic. Heightened blood pressure, hypertension, is a widespread health condition that considerably increases the risk of health complications and diseases. One of the main problems that make the development of a algorithm for generalized blood pressure difficult is the unique morphologies of the patient's biosignals.

Several approaches have been proposed for blood pressure estimation and hypertension detection. The two most popular approaches that are utilized are: feature engineering and deep learning models. The aim behind the former method is to manually define features or automatically extract them so that they correlate with blood pressure. The most significant feature usually is pulse wave metric such as pulse transit time [6] and pulse arrival time. It is a measure of how fast the pulse wave propagates between two different locations on the body.

The latter approach, as well as the one used in this study, attempts to build neural networks capable of learning and identifying complex data patterns that are indicative of blood pressure. This approach lacks the transparency of the previous one, and also generally requires more data to train. However, it doesn't require specialized knowledge and allows the algorithm to use abstract features.

In the domain of cuffless blood pressure estimation from physiological signals, it is commonly accepted that using signals of longer length leads to better results, especially for systolic blood pressure estimation [7]. When working with models for blood pressure estimation, the input signal can have either a fixed or variable length. However, using a fixed window size approach can introduce bias in the dataset because a window with more heartbeats will contain more data. Nevertheless, the fixed window size approach remains a popular choice in many studies because it is computationally efficient and easier to implement [8]. Some studies in this domain focus on continuous cuffless blood pressure estimation, where the goal is to estimate blood pressure continuously over time [9]. Other studies attempt to estimate blood pressure at a single point in time, which is useful in clinical settings where quick measurements are necessary. The choice between continuous and single-point estimation depends on the specific application, the quality and quantity of the dataset, and the desired level of accuracy.

The idea of using hybrid transformer models for estimating blood pressure is the subject of several previous scientific studies. One such study [10] proposes a KD-Informer pretrained on a large high-quality dataset before transferring that knowledge onto the target dataset. This is done in the hopes of integrating prior information of PPG patterns into the model. The results achieved by this approach are an estimation error of $0.02 \pm 5.93$ mmHg for SBP and $0.01 \pm 3.87$ mmHg for DBP [10]. The MLP Mixer is a type of neural network that is heavily inspired from the transformer self-attention mechanism [11]. Based on this architecture a study proposes a MLP-BP method that use both ECG and PPG for blood pressure estimation [12]. The results achieved by this study MAE of 2.13 (2.47) mmHg and SD of 3.07 (3.52) mmHg for DBP and MAE of 3.52 (4.18) mmHg, and a SD of 5.10 (5.87) for SBP [12], meet the highest requirements set by the AAMI [13] and BHS [14] standards.

## 2   Methodology

### 2.1   Objectives

The main goal of this study is to develop a robust blood pressure estimation model which could be used in real life application. To accomplish this goal, we make a hypothesis that the transformer architecture is suitable for this task and may even overtake the results achieved by previous studies with more traditional models. A simple experiment is performed to check the reliability and credibility of our chosen model architectures. While such an analysis might not be necessary, we believe it's prudent because of the architecture's novelty and the overall exploration of the relevance of transformers in biosignal processing. The experiment we focus on is the task of ECG heartbeat classification, owing to the simplicity of the problem and availability of a public dataset.

### 2.2   Evaluation Metrics

In this section we define the metrics used for our models evaluation. For the classification problem used for validation of the architectures credibility:

1. Accuracy is the ratio of corrected predictions to the total number of predictions.
2. Macro F1 score is the harmonic mean of precision and recall calculated for each class separately and then averaged.
3. Weighted F1 score is calculated as the harmonic mean of the precision and recall scores for each class averaged taking into account the classes support.

For the evaluation of the regression models we use the following metrics:

1. Mean Absolute Error, MAE, is the average absolute difference between the predicted and actual values.
2. Mean Squared Error, MSE, is a regression metric representing the average squared difference between the predicted and actual values.
3. Root Mean Squared Error, RMSE, is calculated as the root of the mean squared error.

## 2.3   Framework

The transformer architecture was originally designed for natural language processing. It significantly outperformed the existing architecture in this domain and it's currently considered a state-of-art. Drawing inspiration from the original transformer, several variant of this architecture have been developed for time series processing. Some of the most commonly used models for this task are the time series transformer, the convolutional vision transformer, the temporal fusion transformer and the informer. These models have been thoroughly researched in a previous study of ours [15]. The latter two models are cumbersome in regard to data required for training and computational resources. While the former two and the ones used as part of this study are considered lightweight models and they have more constraint on the type of data they can process. They aren't meant to work with missing data or time series sampled at an irregular interval. The data is prepared with considering these limitation with the hope that the created portable models capable to function in real time on low-spec device.

**Time Series Transformer,** TST, is a variant of the transformer architecture, specifically adapted for time series. It has a relatively simple design when compared to other architectures meant for handling sequential input data, such as the temporal fusion transformer and the informer. This type of model lacks the ability to handle an irregular time series, as well as missing data. Common uses include time series forecasting and anomaly detection. The implementation of the algorithm used for TST can be found at [16] and it has been adapted for regression to suit our needs. A schema of the time series transformer used in this study is shown on Fig. 1.

**Convolutional Vision Transformer,** CvT, is a type of a neural network that employs the convolutional vision transformer blocks as part of its structure. The strength of convolutional layers is their ability to detect local patterns in data. Although this generally applies to two-dimensional data, it can be easily adapted to handle one-dimensional sequential data, such as a time series. The general idea behind this model type is the combination of convolutional and transformer architectures creates a model that can detect both local and global data patterns. The implementation of CvT used in our paper is coded using TensorFlow [17], a popular deep learning framework. A diagram of the model we are using is illustrated on Fig. 2.

It's important to note that even though both of these architectures are hybrid transformers, they lack a decoder, which is necessary for NLP, the problem Transformers were originally designed for.

The Adaptive Moment Estimation, ADAM, optimization algorithm was used for the model training, which is one of the most popular methods.

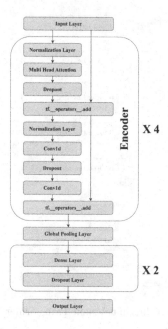

**Fig. 1.** Time Series Transformer Schema

## 2.4 Validation of the Model Architectures Credibility

Both, the time series and the convolutional vision transformers, are fairly new and have not been extensively tested and validated in the context of biosignal processing. Therefore, it is important to conduct an investigation to assess their credibility before they can be reliably used in practical applications. The method used for validating the models credibility is a straightforward comparison with other models.

Heartbeat classification is the process of sorting the individual heartbeats of an ECG. This is done in order to detect arrhythmia or an irregular heart rhythm. AAMI recommends 15 classes for arrhythmia classification, which can be grouped into 5 superclasses: Normal (N) beat, Supraventricular ectopic beat (SVEB), Ventricular ectopic beat (VEB), Fusion beat (F) and Unknown beat (Q) [18]. Arrhythmia detection is vital in the diagnosis and monitoring of many cardiac conditions.

The dataset used in our research is a subset from the PhysioNet MIT BIH [19] dataset. The dataset we are using for our research is the publicly accessible ECG heartbeat categorization dataset, which is available on Kaggle [20]. The dataset overall contains about 100 000 rows, 20 % of which are used for testing. This dataset is already preprocessed into individual heartbeats with a fixed length. An oversampling approach is utilized to overcome the class imbalance present in the dataset. The input is a single ECG heartbeat feed to the network as a time series and the output is one of the superclasses of arrhythmia.

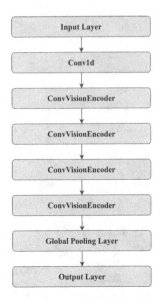

**Fig. 2.** Convolutional Vision Transformer Schema

Table 1 shows a comparison of the results achieved by the different models. Fig 3a and Fig 3b give the normalized confusion matrices obtained by the time series and the convolutional vision transformers, respectively.

**Table 1.** Heartbeat Classification Results Comparison

| Model | Accuracy | Macro F1 Score | Weighted F1 Score |
|-------|----------|----------------|-------------------|
| CNN   | 0.97     | 0.85           | 0.97              |
| TST   | 0.91     | 0.72           | 0.92              |
| CvT   | 0.92     | 0.75           | 0.93              |

The conducted credibility validation demonstrates the suitability of the models for addressing the heartbeat classification problem when presented in the aforementioned format. By inputting an ECG signal and processing it through the network, the models successfully determine the corresponding heartbeat class. It is crucial to emphasize again that the heartbeat classification problem is not the primary focus of our research, but rather a preliminary trial employing a dataset from a similar domain to the problem we are actually focusing to solve. The implementations of the Time Series Transformer and the Convolutional vision Transformer have undergone rigorous testing and validation, leading to the conclusion that these models possess credibility for practical application in biosignal processing.

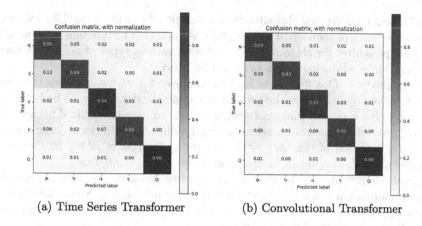

(a) Time Series Transformer        (b) Convolutional Transformer

**Fig. 3.** Normalized Confusion Matrices

## 2.5 Data Collection Procedures

In this study we use a subset of the publicly available MIMIC III dataset [21]. This is a vast dataset that contains a deidentified health information of over 40000 patients. This study is a part of a larger developmental project and a subset of the dataset has been made available in that scope, which we further refined. For our purposes we filtered out the data that contains PPG and arterial blood pressure, ABP. The sampling rate needs to be constant in all of the recordings and in our case it is 125 Hz. Both of the chosen transformer's data can't natively handle missing data. There are multiple ways to handle this problem, such as interpolation of the missing values. However, since the dataset is quite large, the simpler approach and the one used in this study is the exclusion of all signals with missing values, either in the PPG or the ABP part. Another condition for signal exclusion is extremely improbable data. This increases bias and limits the models ability to predict edge cases. However, this is an approach that has been often employed in this type of studies [22]. The ranges we have decided to employ are (40, 120) mmHg for DBP, (60, 200) mmHg for SBP, and (20, 80) mmHg for SBP-DBP, that have been determined considering the domain in question.

## 2.6 Data Analysis

Several steps are taken to construct the dataset that ensure the quality of the data for our model's training and testing. This process was inspired by another study [23] that was conducted in the scope of the larger project "SP4LIFE". It should be noted that proper processing is necessary because of the problem's complexity.

1. Data Cleaning. Noise caused by motion artifacts and baseline wandering may be present in the PPG signal, which can impact the accuracy of subsequent

methods. To overcome this situation, a function from the Neurokit2 toolbox for processing physiological signals [24] was applied. The function implements a bandpass filter for cleaning the signal from various types of noise that may be encountered during the data collection process.

2. Normalization. This technique is often employed for various analysis of information. Normalization is the process of scaling the data range to be between 0 and 1. It's a type of min max scaling and in our case it's utilized because of sensitivity of the neural networks to larger values. An additional justification for the inclusion of this step is the variability in data recorded by different sensors, which may result in divergent scales across the dataset.

3. Signal Alignment. The process of aligning two different signals is necessary, since more than often the recordings are unsynchronised. Signal alignment can be achieved through various approaches [25,26], and among them, we select the cross-correlation algorithm.

4. Pearson Correlation. As stated, the used dataset is vast, and the signal quality varies. The Pearson coefficient is a statistical measure that describes the linear correlation between two variables. Only if the Pearson coefficient of the PPG-ABP pair exceeds the threshold of 0.85, the signals are taken into consideration. This is done in order to exclude the signals that contain noise and interference.

5. Segmentation. Determining the length of the signals we use to build our model is an important step. Although we recognize that a larger window size will likely lead to better results, we choose to use a smaller one to lower computational cost. Multiple studies have had success using short-term PPG for blood pressure estimation [27].

## 3    Results

The dataset contains information from 71 patients, and a division is made to create distinct training, validation, and testing datasets. To ensure data independence, the samples from one patient are present in only one of the datasets. The testing and validation dataset contain 10 patients each. In the literature, it's usually suggested to use at least 10 patients for testing. It's also recommended these patients to be of different ages, genders, and a afflicted by variety of conditions, such as pregnancy and diabetes. However, this isn't something we can take into consideration since the data is anonymized and the patients can only be differentiated by a patient identifier.

The aforementioned time series transformer (TST) and a convolutional vision transformer (CvT) were built and evaluated on these datasets. The TST model consist of 4 encoder blocks stacked on top of each other, each contains a self attention mechanism with 4 heads that allow the model to focus on different parts of the input. Each head has a length of 8 to help it learn the patterns in the data. The CvT model on the other hand, has 4 encoder blocks and a kernel size of 8. All of these values were determined empirically, through series of experiments. Table 2 shows a comparison of the results achieved by our models.

**Table 2.** BP Estimation Results Comparison

| Model | SBP (mmHg) | | | DBP (mmHg) | | |
|---|---|---|---|---|---|---|
| | MAE | MSE | RMSE | MAE | MSE | RMSE |
| TST | 17.07 | 408.02 | 20.2 | 8.79 | 121.87 | 11.04 |
| CvT | 16.66 | 408.16 | 20.20 | 9.25 | 123.76 | 11.12 |

Figure 4 gives the per subject results for both SBP and DBP using the TST model. This analysis shows that some subjects are accurately estimated, while others are significantly off the mark. The table also highlights the bias introduced by an uneven number of samples per patient.

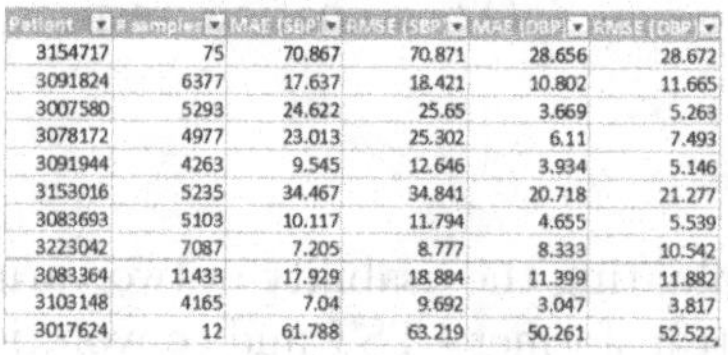

| Patient | # samples | MAE (SBP) | RMSE (SBP) | MAE (DBP) | RMSE (DBP) |
|---|---|---|---|---|---|
| 3154717 | 75 | 70.867 | 70.871 | 28.656 | 28.672 |
| 3091824 | 6377 | 17.637 | 18.421 | 10.802 | 11.665 |
| 3007580 | 5293 | 24.622 | 25.65 | 3.669 | 5.263 |
| 3078172 | 4977 | 23.013 | 25.302 | 6.11 | 7.493 |
| 3091944 | 4263 | 9.545 | 12.646 | 3.934 | 5.146 |
| 3153016 | 5235 | 34.467 | 34.841 | 20.718 | 21.277 |
| 3083693 | 5103 | 10.117 | 11.794 | 4.655 | 5.539 |
| 3223042 | 7087 | 7.205 | 8.777 | 8.333 | 10.542 |
| 3083364 | 11433 | 17.929 | 18.884 | 11.399 | 11.882 |
| 3103148 | 4165 | 7.04 | 9.692 | 3.047 | 3.817 |
| 3017624 | 12 | 61.788 | 63.219 | 50.261 | 52.522 |

**Fig. 4.** Per Subject Results - TST

## 3.1   Discussion

Despite the initial expectation that the built TST and CvT models may be used for blood pressure estimation, the obtained results are not entirely satisfactory, as shown on Table 2. The BHS and AAMI standards set criteria for grading of blood pressure measuring devices. The minimum passing grade for the AAMI standards is achieved when the mean error is within 10 mmHg for SBP and 5 mmHg for DBP, and the standard deviation is 10 mmHg for SBP and 8 mmHg for DBP [13]. Numerous studies have investigated the connection between ABP and PPG, consistently highlighting PPG as a reliable indicator of ABP. However, these models did not effectively incorporate this connection. Many factors influence the results.

A critical point is the data used for building and evaluating the models. The dataset is obtained from MIMIC III which is the typical for this type of problems and the preprocessing steps undertaken are similar to those of previous studies. A greater dataset with higher quality signals may allow this type of models to improve. Another issue that negatively effects the experiments outcome is the lack of implementation of an optimization process. During the model training process, no hyperparameter tuning was conducted, and the values for the hyperparameters were chosen using heuristics. As future work, employing a comprehensive optimization process that performs a systematic approach for tuning the models hyperparameters will undoubtedly lead to better results. One potential explanation for why the explored transformer architectures are unsuitable for this type of problem is their limited capacity to generalize and accommodate the distinctive morphologies of various patients. Another concern is whether models built solely on PPG data is sufficient. Certain studies have achieved results with this approach, however typically for this type of problem ECG data is used in conjecture with PPG. It's also notable to mention that these architectures are relatively simple in regard to processing time series. Other more sophisticated transformer frameworks, such as the temporal fusion and informer, may lead to better results.

A practical application of the elaborated models is developing a multisensor chest-patch (ECG, PPG, Graphene). Given the signals, the device will have the processing power to provide continuous measurement and real-time calculation of the Heart rate and Respiratory rate and use the signals to feed the developed models for Blood pressure and SpO2 estimation, respectively [28].

## 4    Conclusion

This study focuses on evaluating the usability of two variants of the transformer architecture (time series transformer - TST and convolutional vision transformer - CvT) for processing biological signals. To assess their performance, a validation check is conducted on a different solved biosignal problem within the same domain, where the used architectures demonstrate satisfactory results. Subsequently, similar models are constructed to estimate blood pressure using a PPG signal. A subset of Physionet MIMIC III dataset is prepared and several models are built. There is no significant difference between the models's performances - the obtained MAE results are: TST (SBP: 17.07 mmHg; DBP: 8.79 mmHg) and CvT (SBP: 16.66 mmHg; DBP: 9.25 mmHg). The other metrics are given in Table 2. As far as we are aware, there are no existing studies on the application of TST and CvT models specifically for blood pressure estimation. Therefore, we are unable to compare our results with those of other TST and CvT models in this context.

The selected transformer architectures, namely the time series and convolutional vision transformers, did not obtain satisfactory results for blood pressure estimation. For accurate estimation, the Mean Absolute Error (MAE) of SBP and DBP should be less than 5 mmHg, according the AAMI standards. However, it is important to note that further experimentation is necessary before concluding that these models are unsuitable for this task.

As a future work, fine-tuning the hyperparameters of these models and enlarging the dataset could potentially improve the results significantly, to achieve more accurate and reliable blood pressure estimation.

**Acknowledgment.** This paper has been written thanks to the support of the "Smart Patch for Life Support Systems" - NATO project G5825 SP4LIFE and by the National project IBS4LIFE of Faculty of Computer Science and Engineering, at Ss. Cyril and Methodius University in Skopje.

# References

1. Vaswani, A., et al.: Attention is all you need. In: Advances in Neural Information Processing Systems, vol. 30 (2017)
2. Escabí, M.A.: Biosignal processing. In: Introduction to Biomedical Engineering. Elsevier, pp. 549–625 (2005)
3. Berkaya, S.K., Uysal, A.K., Gunal, E.S., Ergin, S., Gunal, S., Gulmezoglu, M.B.: A survey on ECG analysis. Biomed. Signal Process. Control **43**, 216–235 (2018)
4. Cheriyedath, S.: Photoplethysmography (PPG). news-medical (2019). https://www.news-medical.net/health/Photoplethysmography-(PPG).aspx,. Accessed 30 July 2023
5. High blood pressure causes and risk factors (2022). https://www.nhlbi.nih.gov/health/high-blood-pressure/causes,. Accessed 12 May 2023
6. Smith, R.P., Argod, J., Pépin, J.-L., Lévy, P.A.: Pulse transit time: an appraisal of potential clinical applications. Thorax **54**(5), 452–457 (1999)
7. Liu, M., Po, L.-M., Fu, H.: Cuffless blood pressure estimation based on photoplethysmography signal and its second derivative. Int. J. Comput. Theory Eng. **9**(3), 202 (2017)
8. Mousavi, S.S., Firouzmand, M., Charmi, M., Hemmati, M., Moghadam, M., Ghorbani, Y.: Blood pressure estimation from appropriate and inappropriate PPG signals using a whole-based method. Biomed. Signal Process. Control **47**, 196–206 (2019). https://www.sciencedirect.com/science/article/pii/S1746809418302209
9. Kachuee, M., Kiani, M.M., Mohammadzade, H., Shabany, M.: Cuffless blood pressure estimation algorithms for continuous health-care monitoring. IEEE Trans. Biomed. Eng. **64**(4), 859–869 (2016)
10. Ma, C., et al.: KD-informer: a cuff-less continuous blood pressure waveform estimation approach based on single photoplethysmography. IEEE J. Biomed. Health Inform. **27**(5), 2219–2230 (2023)
11. Tolstikhin, I.O., et al.: MLP-mixer: an all-MLP architecture for vision. In: Advances in Neural Information Processing Systems, vol. 34, pp. 24261–24272 (2021)
12. Huang, B., Chen, W., Lin, C.-L., Juang, C.-F., Wang, J.: MLP-BP: a novel framework for Cuffless blood pressure measurement with PPG and ECG signals based on MLP-mixer neural networks. Biomed. Signal Process. Control **73**, 103404 (2022)
13. White, W.B., et al.: National standard for measurement of resting and ambulatory blood pressures with automated sphygmomanometers. Hypertension **21**(4), 504–509 (1993)
14. O'Brien, E., et al.: The British hypertension society protocol for the evaluation of blood pressure measuring devices. J. Hypertens. **11**(Suppl 2), S43–S62 (1993)
15. Kuzmanov, I., Ackovska, N., Madevska Bogadnova, A.: Transformer models for processing biological signal (2023)
16. Ntakouris, T.: Timeseries classification with a transformer model (2021). https://keras.io/examples/timeseries/timeseries_transformer_classification/. Accessed 13 May 2023
17. M. Abadi, A. Agarwal, P. Barham, E. Brevdo, Z. Chen, C. Citro, G. S. Corrado, A. Davis, J. Dean, M. Devin, S. Ghemawat, I. Goodfellow, A. Harp, G. Irving, M. Isard, Y. Jia, R. Jozefowicz, L. Kaiser, M. Kudlur, J. Levenberg, D. Mané, R. Monga, S. Moore, D. Murray, C. Olah, M. Schuster, J. Shlens, B. Steiner, I. Sutskever, K. Talwar, P. Tucker, V. Vanhoucke, V. Vasudevan, F. Viégas, O. Vinyals, P. Warden, M. Wattenberg, M. Wicke, Y. Yu, X. Zheng, "TensorFlow:

Large-scale machine learning on heterogeneous systems," 2015, software available from tensorflow.org. [Online]. Available: https://www.tensorflow.org/

18. Luz, E.J.D.S., Schwartz, W.R., Cámara-Chávez, G., Menotti, D.: ECG-based heartbeat classification for arrhythmia detection: a survey. Comput. Methods Prog. Biomed. **127**, 144–164 (2016). https://www.sciencedirect.com/science/article/pii/S0169260715003314

19. Moody, G.B., Mark, R.G.: The impact of the MIT-BIH arrhythmia database. IEEE Eng. Med. Biol. Mag. **20**(3), 45–50 (2001)

20. Fazeli, S.: ECG heartbeat categorization dataset (2022). https://www.kaggle.com/datasets/shayanfazeli/heartbeat?datasetId=2941. Accessed 12 May 2023

21. Johnson, A.E., et al.: MIMIC-III, a freely accessible critical care database. Sci. Data **3**(1), 1–9 (2016)

22. Baker, S., Xiang, W., Atkinson, I.: A hybrid neural network for continuous and non-invasive estimation of blood pressure from raw electrocardiogram and photoplethysmogram waveforms. Comput. Methods Prog. Biomed. **207**, 106191 (2021)

23. Mladenovska, T., Bogdanova, A.M., Kostoska, M., Koteska, B., Ackovska, N.: Estimation of blood pressure from arterial blood pressure using PPG signals (2023)

24. Makowski, D., et al.: NeuroKit2: a python toolbox for neurophysiological signal processing. Behav. Res. Methods **53**(4), 1689–1696 (2021). https://doi.org/10.3758%2Fs13428-020-01516-y

25. Shin, H., Min, S.D.: Feasibility study for the non-invasive blood pressure estimation based on PPG morphology: normotensive subject study. Biomed. Eng. Online **16**, 1–14 (2017)

26. Xing, X., Sun, M.: Optical blood pressure estimation with photoplethysmography and FFT-based neural networks. Biomed. Opt. Express **7**(8), 3007–3020 (2016)

27. Chowdhury, M.H., et al.: Estimating blood pressure from the photoplethysmogram signal and demographic features using machine learning techniques. Sensors **20**(11), 3127 (2020). https://www.mdpi.com/1424-8220/20/11/3127

28. Lehocki, F., et al.: Smartpatch for victims management in emergency telemedicine. In: 2021 13th International Conference on Measurement, pp. 146–149. IEEE (2021)

# The Curious Case of Randomness in Deep Learning Models for Heartbeat Classification

Marjan Gusev[1,2]([✉]) [iD], Stojancho Tudjarski[1,2] [iD], Aleksandar Stankovski[1,2] [iD], and Mile Jovanov[1] [iD]

[1] Faculty of Computer Science and Engineering, Sts Cyril and Methodius University in Skopje, Skopje, North Macedonia
{marjan.gushev,mile.jovanov}@finki.ukim.mk
[2] Innovation Dooel, Skopje, North Macedonia
{stojancho.tudjarski,aleksandar.stankovski}@innovation.com.mk

**Abstract.** The research hypothesis in this study is that different random number generator seeds using 1D Convolutional Neural Networks impact the performance results by more than 15% on the heartbeat classification performance. Furthermore, we address a research question to evaluate the impact level of random values in the initialization of model parameters experimenting on the classification of ventricular heartbeats in electrocardiogram training and evaluating models with various feature sets based on the width of the measured samples surrounding a heartbeat location. Specific test cases consist of differently selected initial neural network parameters guided by manually selected random number seeds while preserving the rest of the training environment and hyperparameters. We examine the influence of the random number seed on the model's learning dynamics and ultimate F1 score on the performance of the testing dataset and conclude fluctuations resulting in 24.61% root mean square error from the average. Furthermore, we conclude that optimizing the validation in the training process does not optimize the performance in the testing. The research results contribute a novel viewpoint to the field, paving the way for more efficient and accurate heartbeat classification systems and improving diagnostic and prognostic performance in cardiac health.

**Keywords:** ECG · Heartbeat classification · Convolutional Neural Networks · Deep Learning

## 1 Introduction

Electrocardiogram (ECG) is the electrical representation of heart activity, and ECG signal analysis is essential for diagnosing and monitoring various cardiovascular disease conditions. The development of fast, reliable, and accurate automatic ECG signal classification models in healthcare and biomedical engineering is of the utmost importance to prevent severe heart damage.

M. Mihova and M. Jovanov (Eds.): ICT Innovations 2023, CCIS 1991, pp. 59–75, 2024.
https://doi.org/10.1007/978-3-031-54321-0_5

Ventricular (V) beats are heartbeats originating from the ventricles, the heart's lower chambers, as opposed to the sinoatrial (SA) node, the heart's natural pacemaker in the right atrium. An increased number of V beats indicate an increased risk of health complications, such as coronary artery disease, heart failure, or onset of heart attacks. Due to the different electrical pathways to activate the heart, V beats on ECG recordings typically have an abnormally wide and irregular shape [11]. Figure 1 presents the main characteristic points (Q, R, and S), indicating the R point as the heartbeat complex peak.

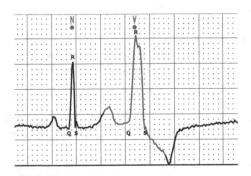

**Fig. 1.** Normal (N) and Ventricular (V) heartbeats differ in their shape. Q, R, and S characteristic points determine the heartbeat features.

This study focuses on V-beat classification models using 1D Convolutional Neural Networks (CNNs) as a Deep Learning (DL) model. The application of CNNs in ECG processing and classification is the subject of research by several authors [5, 6, 13, 14, 17–19]. Despite their implementation success, the challenge of developing a more accurate system still motivates the researchers. In this study, we investigate the DL frameworks and the impact of randomness in the training process.

A source of various shapes and multiple forms of V beats in our experimental methodology is the well-known MIT-BIH Arrhythmia benchmark database [10]. To promote a reliable and generalized model performance, we divide our data into training and testing subsets using a widely accepted patient-oriented approach proposed by DeChazal [2]. This method ensures a roughly equal distribution of heartbeat types across subsets, thereby addressing concerns regarding overfitting and enhancing model generalizability.

Our *research hypothesis* is that randomly selected initialization values of the initial CNN parameters affect the model performance by generating more than 15% root mean square error (RMSE). The *research question* evaluates this influence.

The preliminary research shows that these initialization values can significantly affect the model's learning dynamics and subsequent performance. Exposing the effect of randomly generated initialization values adds dimension to the feature space of experiments designed to identify the best-performing meta parameters. These developments could significantly improve diagnostic performance, prognostic abilities, and cardiac health outcomes. Therefore, they contribute to the evolution of heartbeat classification models.

Related work analyzes similar studies in Sect. 2. Methods are described in Sect. 3 with details on the randomness concept and random seeds in training DL NNs, the dataset, feature engineering, and experimental and evaluation methodology. Section 4 presents the experimental results of our experiments, and their evaluation is discussed in Sect. 5. Section 6 elaborates on the conclusions and directions for future work.

## 2    Related Work

Several authors target the impact of the initial parameter values for DL NNs and evaluate how the randomness impacts the model development, similar to our research question. Wang et al. [16] tested the DL library via effective model generation since training and model development depend on randomness and floating-point computing deviations. The DL libraries are tested as sufficiently as possible by exploring unused library code or different usage ways of library code and do not evaluate the model performance fluctuations presented in this paper.

Computer random number generators (RNGs) are fundamentally deterministic. The algorithm starting point is the RNG initial state (seed) and determines the entire sequence of numbers. The operation of the RNG is dependent on its ability to generate a random number sequence without discernible patterns or repetition, which is essential for the initialization of ML algorithm parameters. The "randomness" of these numbers is crucial to ensuring that the deterministic nature of the computer's operations does not skew or bias them and draws attention to reproducibility as an essential aspect. Marrone et al. [9] assess the reproducibility to determine if the resulting model produces good or bad outcomes because of luckier or blunter environmental training conditions. Although this is a non-trivial DL problem, the authors conclude that training and optimization phases strongly rely on stochastic procedures and that the use of some heuristic considerations (mainly speculative) to reduce the required computational effort tends to introduce non-deterministic behavior with a direct impact on the results of biomedical image processing applications relying on deep learning approaches. The authors do not evaluate the performance fluctuation levels, which is our focus.

Hooper et al. [4] evaluate the impact of tooling on different training architectures and datasets and discuss random initialization as one of the algorithmic sources of randomness mainly caused by the weights from sampling random distribution. Authors observe an insignificant variance in accuracy with less than 1% standard deviation on obtained results. We found these fluctuations are much higher in the evaluated use case with an imbalanced dataset.

A significant performance impact of particular algorithms is observed by a proper selection of an RNG seed, especially those involving stochastic processes.

Different seed values will produce different initial weights, which may result in distinct network behaviors, especially in models with non-convex loss landscapes and complex structures. Despite the undeniable importance of the seed value, it is essential to note that excessive reliance on a single seed value or an optimization over seed values may result in overfitting or misleading results. The model performance should not be overly dependent on the seed value, and a more profound issue must be addressed if it is. Glorot and Bengio [3] evaluate why standard gradient descent from random initialization is doing so poorly with DL NNs and help design better algorithms in the future. They found that the top hidden layer is driven into saturation due to the logistic sigmoid activation unsuited for DL NNs with random initialization. In addition, the saturated units can move out of saturation by themselves, albeit slowly and explaining the plateaus sometimes seen when training NNs. The authors investigate the reasons without a deeper evaluation of the problem size addressed in this paper.

Choice of tooling can introduce randomness to DL NN training, and Zhuang et al. [20] found that accuracy is not noticeably impacted. Although, they conclude that model performance on certain parts of the data distribution is far more sensitive to randomness, similar to our conclusions which are documented with significant performance fluctuations up to 25%.

The initial parameter selection for a Machine Learning (ML) model substantially impacts the training procedure and the model's final performance. This is due to the nature of the optimization algorithms used in ML, which are typically iterative methods that gradually adjust model parameters to minimize a loss function - a measure of the difference between the model's predictions and the actual data. Training a model can be compared to traversing a landscape of potential parameters shaped by the loss function. This landscape consists of numerous valleys and craters, as various local minima (sets of parameters for which the loss function reaches a local minimum.) The training objective is to identify the most bottomless pit (the global minimum) as the set of parameters for which the loss function reaches its absolute minimum value, as discussed by Silva et al. [15]. Multiple local extremes are illustrated in Fig. 2, similar to the conclusions in this paper.

The Stochastic Gradient Descent (SGD) algorithm mathematically models the walk through this pathway. The initial selection of parameters determines the starting point of the path. If the starting point is close to a shallow local minimum, the optimization algorithm may become stalled and be unable to locate deeper, superior minima. This can result in subpar model performance. When the starting point is close to the global minimum or a deep local minimum, the algorithm has a greater chance of finding a good set of parameters resulting in superior model performance.

Typically, random initialization of model parameters is used to address this issue. The likelihood of discovering an excellent local extreme or even the global one is increased if the algorithm starts with different random seeds. This method, however, increases computational requirements. Advanced techniques, such as sophisticated initialization strategies or momentum-based optimization methods,

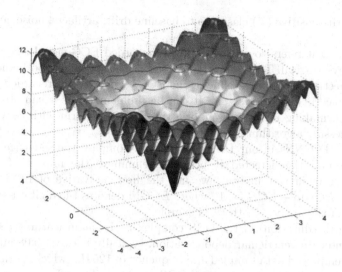

**Fig. 2.** Local and global minimums [15]

can aid in navigating the parameter landscape more efficiently and effectively, frequently resulting in enhanced model performance.

The influence of chosen initial model parameters is analyzed by Liu et al. [8]. They observe that their SGD algorithm may converge to the closest local optima close to the initial point in the parameters space. Even though this seems natural, considering how SGD works, the value of his work is that he has shown practically how a deliberately chosen bad initial point in the parameters space forces the model to finalize its fitting process to a bad local optimum. Li et al. [7] provide a well-elaborated analysis of the behavior of SGD when used to train overparameterized NNs. They proved that SGD produces valuable models with sufficiently good generalization behavior when applied over initial randomly chosen parameters as an excellent way for wanted NN training. The rationality behind his conclusion is that the randomly initialized parameters produce an even distribution of negative and positive weights. It also lowers changes for appearing outputs at the networks' layers, which are significantly big or small. This leads to the formation of smooth, easily generalizable networks. In this paper, we have not analyzed the SGD model in detail, but experimental results confirm the existence of local optima and large performance fluctuations.

## 3   Methods

### 3.1   Use Case: V Versus NonV Classification of Heartbeats

The use case in our experiments is the classification of the V beats versus NonV beats. Classifying a V beat is usually determined by its different shape in the signal processing algorithms. It is difficult to distinguish if a particular heartbeat belongs to the positive (V) class due to baseline drift, artifacts, noise, and other side effects.

Baseline drift refers to variations in the recorded ECG signal that are slow and non-physiological. This typically manifests as a slow and smooth undulation of the isoelectric line (the baseline of the ECG signal) and is regarded as noise or interference in the ECG signal. It can be problematic as it can obscure the ECG signal, making it hard to analyze the ECG waveform with conventional signal processing algorithms.

We use 1D CNNs in our DL modeling and compare the sequence of ECG samples around the detected R peak. To eliminate the impact of different isoelectric start in the analysis, we use the first derivative of ECG samples $dECG[i] = ECG[i] - ECG[i-1]$, where $ECG[i]$ denotes the digital representation value of the analyzed $i$-th ECG sample.

The width (duration) of the QRS complex is the main feature used in our model to indicate ventricular depolarization, typically between 0.08 and 0.12 s. Our study employed data sampled at a frequency of 125 Hz (which results in an 8 ms time duration between subsequent ECG samples). This effectively means the typical QRS temporal duration is between 10 to 15 discrete samples. Our goal was to incorporate additional electrophysiological data, specifically the preceding P wave, which represents atrial depolarization, and the subsequent T wave, which represents ventricular repolarization, to improve the diagnostic precision of ventricular beat detection since they define the irregular shape of the V beat much better.

### 3.2   1D CNN Architecture and Training Hyperparameters

Our preferred model development and experimental environment is the TensorFlow v.2.10.1 Deep Learning framework. The following API generates the model:

```
Layer type    Output Shape   Param #
=============================================
Conv1D  (None, 20, 32)    128
Conv1D  (None, 18, 128)   12416
MaxPooling 1D    (None, 9, 128)   0
Dropout (None, 9, 128)    0
Flatten (None, 1152)   0
Dense   (None, 64)    73792
Dense   (None, 64)    4160
Dropout (None, 64)    0
Dense   (None, 2)    130
=============================================
Total params: 90,626
Trainable params: 90,626
Non-trainable params: 0
```

The training hyperparameters are listed in Table 1. We implemented an early stopping training strategy, wherein the training procedure is halted when the

**Table 1.** 1-D CNN training hyperparameters

| Name | Value |
| --- | --- |
| Loss function | Binary Cross-entropy [12] |
| Learning rate | 0.001 |
| Optimizer | Adam [1] |
| beta-1 | 0.9 |
| beta-2 | 0.999 |
| epsilon | 1e−07 |
| Maximum training epochs | 50 |

validation accuracy declines observing convergence and decline of loss function. This resulted in the completion of between 19 and 25 training epochs.

### 3.3 Experiments

We conducted a series of experiments in which we varied the number of samples included in the feature set. The length of the sample intervals varies between 23 to 47, with the sample split left and right of the R-peak approximately 33% at the left and 66% at the right, with motivation:

– To capture a significant portion of the entire P wave, depending on the signal's particular temporal characteristics, and
– To capture a significant portion of the entire T wave, which has a longer duration and higher variability than the P wave and QRS complex. By incorporating a broader range of samples, we aimed to account for the inherent heterogeneity of T wave characteristics across the population, thereby improving the generalizability and robustness of the ventricular beat detection models.

Our investigation employed feature sets containing between seven and 15 samples left of the R-peak and between 15 and 31 right of the R-peak, with the constraint of having approximately twice as samples at the right as those at the left. We initialize the model with distinct values for the RNG seeds to evaluate the effect of arbitrarily chosen initial CNN parameter values.

### 3.4 Dataset

The MIT-BIH Arrhythmia Database (MITDB) is a benchmark database widely recognized and used for arrhythmia detection algorithms in biomedical signal processing and machine learning. The database is maintained by PhysioNet, an online resource for complex physiological signals, and is accessible to the public.

DeChazal's method was utilized to determine train and test subsets from the MITDB. His split is a patient-centric method for splitting training and testing sets. The ECG records in the MITDB are separated into two groups: the training dataset DS1 and the testing dataset DS2, each containing 22 records from different patients, without overlapping, such that an ECG from patients in the training dataset does not belong to the testing dataset. The DS1 dataset contains 51021 heartbeats and DS2 49712, such that 3788 are ventricular beats in DS1 and 3220 in DS2. The class ratio is 7.22% for DS1 and 6.48% for DS2, representing a significant imbalance for classifying ventricular beats in the classification task.

The primary goal of the DeChazal split is to divide the database so that each patient's information is contained in either the training or the test set, but not both. This method ensures that the model is trained and tested on various patients, thereby assessing the model's ability to generalize to patients who have not been observed. This is especially crucial in healthcare applications, where patient-specific characteristics can vary substantially.

The DeChazal split ensures that the distribution of different heartbeats between the training and testing sets is roughly comparable. This helps prevent bias in the model's predictions caused by an imbalanced dataset. Consequently, this split method enables a more precise evaluation of the performance of heartbeat classification models.

### 3.5   Evaluation

Binary classification problems involve predicting positive and negative possible outcomes. **Accuracy** $ACC$ is the most intuitive performance measure calculated as the ratio of correct predictions to the total number of detections. It works fine for balanced sets, and in our case, we deal with a highly imbalanced dataset with less than 7% of V beats versus all heartbeats.

Evaluating the performance of a binary classification model for imbalanced datasets, we focus on the following metrics:

- **Recall (Sensitivity, True Positive Rate)** $SEN$ is the ratio of correct positive predictions to the total actual positives and evaluates the classifier's completeness. Low sensitivity indicates a high number of false negatives.
- **Precision (Positive Predictive Value)** $PPV$ is the ratio of correct positive predictions to the total predicted positives and evaluates the classifier's exactness. Low precision indicates a high number of false positives.
- **F1 Score** $F1 = 2*((PPV*SEN)/(PPV+SEN))$ is the harmonic mean of the sensitivity and positive predictive value and conveys their balance. This is the performance measure to evaluate in the case of imbalanced datasets, such as ours.

To evaluate the differences between results obtained from different test cases, we provide the minimum, maximum, and mean values and the standard deviation of the results. The relative range value is calculated as a ratio of the difference between the maximal and minimal F1 score versus the average value. Furthermore, we express the relative standard deviation.

## 4   Results

The results will be presented in a tabular form to express the fluctuations for achieved F1 scores executed on different test cases. Each row specifies a different sample interval width (WIDTH) with results achieved by distinct RNG seeds. Besides the minimum (MIN), maximum (MAX), average (AVG), and standard deviation (STD) in absolute F1 score value (in %), we present the relative range (RANGE) and relative standard deviation (RSD), as the ratio between the range of values versus the actual mean F1 value. Table 2 presents the results validating the DS1 dataset within the training using the De Chazal DS1/DS2 dataset split, while Table 3 presents the results testing the DS2 dataset.

**Table 2.** F1 score statistics for validating the DS1 dataset using the De Chazal DS1/DS2 inter-patient data split for test cases determined by sample interval widths from 23 to 47 and applying different RNG seeds

| WIDTH | MIN | MAX | AVG | STD | RANGE | RSD |
|---|---|---|---|---|---|---|
| 23 | 89.45 | 96.13 | 92.67 | 2.25 | 7.21 | 2.43 |
| 24 | 89.46 | 94.79 | 92.39 | 1.93 | 5.77 | 2.09 |
| 25 | 89.45 | 94.79 | 92.43 | 1.89 | 5.77 | 2.05 |
| 26 | 89.47 | 94.61 | 92.35 | 1.80 | 5.56 | 1.94 |
| 27 | 89.47 | 96.08 | 92.70 | 2.21 | 7.13 | 2.38 |
| 28 | 89.47 | 94.82 | 92.40 | 1.94 | 5.80 | 2.10 |
| 29 | 89.43 | 96.13 | 92.59 | 2.19 | 7.24 | 2.36 |
| 30 | 89.48 | 95.98 | 92.67 | 2.22 | 7.01 | 2.40 |
| 31 | 89.46 | 94.79 | 92.42 | 1.90 | 5.77 | 2.05 |
| 32 | 89.48 | 94.80 | 92.39 | 1.92 | 5.76 | 2.08 |
| 33 | 89.47 | 96.11 | 92.59 | 2.18 | 7.18 | 2.36 |
| 34 | 89.43 | 96.07 | 92.67 | 2.25 | 7.16 | 2.43 |
| 35 | 89.45 | 94.81 | 92.40 | 1.93 | 5.81 | 2.09 |
| 36 | 89.47 | 94.66 | 92.33 | 1.82 | 5.63 | 1.97 |
| 37 | 89.48 | 96.17 | 92.68 | 2.25 | 7.22 | 2.43 |
| 38 | 89.48 | 96.13 | 92.69 | 2.25 | 7.18 | 2.42 |
| 39 | 89.45 | 94.79 | 92.39 | 1.93 | 5.77 | 2.09 |
| 40 | 89.47 | 94.64 | 92.36 | 1.80 | 5.60 | 1.94 |
| 41 | 89.45 | 96.11 | 92.67 | 2.25 | 7.18 | 2.43 |
| 42 | 89.48 | 94.65 | 92.35 | 1.80 | 5.61 | 1.94 |
| 43 | 89.47 | 96.05 | 92.59 | 2.16 | 7.11 | 2.34 |
| 44 | 89.49 | 94.81 | 92.39 | 1.92 | 5.76 | 2.08 |
| 45 | 89.47 | 94.76 | 92.42 | 1.89 | 5.73 | 2.05 |
| 46 | 89.46 | 94.59 | 92.32 | 1.82 | 5.56 | 1.97 |
| 47 | 89.44 | 94.79 | 92.52 | 1.89 | 5.78 | 2.04 |

The relative range values for validating the DS1 dataset vary from 5.56% to 7.24% and the relative standard deviation from 1.94% to 2.43%. Testing the DS2 dataset, the relative range values vary from 15.51% to 16.68%, and relative standard deviation from 5.19% up to 5.50%. These fluctuations in testing performance are significantly more than validation performance.

An illustration of F1 score performance results is presented in Table 4 for the test case specifying different RNG seeds for the sample interval width of 25. Note that the width sample interval of 25 is obtained with eight samples to the left of the R-peak and sixteen samples to the right.

**Table 3.** F1 score statistics for testing the DS2 dataset using the De Chazal DS1/DS2 inter-patient data split for test cases determined by sample interval widths from 23 to 47 and applying different RNG seeds

| WIDTH | MIN | MAX | AVG | STD | RANGE | RSD |
|-------|-------|-------|-------|------|-------|------|
| 23 | 67.22 | 78.80 | 72.93 | 3.83 | 15.88 | 5.26 |
| 24 | 67.26 | 78.81 | 73.30 | 3.84 | 15.75 | 5.24 |
| 25 | 67.25 | 79.49 | 73.37 | 3.97 | 16.68 | 5.42 |
| 26 | 67.23 | 79.34 | 73.81 | 4.05 | 16.41 | 5.49 |
| 27 | 67.28 | 79.47 | 73.06 | 3.96 | 16.68 | 5.43 |
| 28 | 67.27 | 78.70 | 73.28 | 3.81 | 15.60 | 5.20 |
| 29 | 67.24 | 78.71 | 73.45 | 3.93 | 15.61 | 5.35 |
| 30 | 67.24 | 78.84 | 72.88 | 3.84 | 15.92 | 5.27 |
| 31 | 67.24 | 79.27 | 73.33 | 3.97 | 16.41 | 5.41 |
| 32 | 67.24 | 78.71 | 73.26 | 3.82 | 15.65 | 5.22 |
| 33 | 67.33 | 78.77 | 73.48 | 3.92 | 15.56 | 5.34 |
| 34 | 67.27 | 78.78 | 72.97 | 3.79 | 15.78 | 5.19 |
| 35 | 67.20 | 78.71 | 73.23 | 3.84 | 15.72 | 5.25 |
| 36 | 67.31 | 78.75 | 73.71 | 3.93 | 15.52 | 5.33 |
| 37 | 67.27 | 78.71 | 72.94 | 3.81 | 15.68 | 5.22 |
| 38 | 67.24 | 78.60 | 72.91 | 3.78 | 15.58 | 5.19 |
| 39 | 67.26 | 78.72 | 73.29 | 3.85 | 15.64 | 5.25 |
| 40 | 67.31 | 79.39 | 73.80 | 4.04 | 16.38 | 5.47 |
| 41 | 67.29 | 78.68 | 72.92 | 3.81 | 15.61 | 5.23 |
| 42 | 67.25 | 79.28 | 73.81 | 4.06 | 16.30 | 5.50 |
| 43 | 67.28 | 78.76 | 73.41 | 3.96 | 15.64 | 5.39 |
| 44 | 67.23 | 78.80 | 73.27 | 3.87 | 15.79 | 5.29 |
| 45 | 67.20 | 79.16 | 73.26 | 3.92 | 16.32 | 5.35 |
| 46 | 67.27 | 78.70 | 73.72 | 3.90 | 15.51 | 5.29 |
| 47 | 67.22 | 78.77 | 73.26 | 3.86 | 15.76 | 5.27 |

# 5    Discussion

The variety of achieved results in our experiment was presented in Tables 2 and 3 as a summary covering all test cases.

## 5.1    Impact of RNG Seeds

Figure 3 presents more details on the achieved results of the test case with a sample interval width of 25 for the first 2000 different RNG seeds. We observe some isolated specific results when the random seed has corrupted the training process and led to the performance of a 25% F1 score.

**Table 4.** F1 scores from the models trained with a sample interval width of 25 ECG samples per heartbeat and different RNG seeds.

| Seed | Validation | | | Testing | | |
|------|-------|-------|-------|-------|-------|-------|
|      | SEN   | PPV   | F1    | SEN   | PPV   | F1    |
| 1    | 91.29 | 95.68 | 93.43 | 83.49 | 62.23 | 71.31 |
| 2    | 93.10 | 96.53 | 94.79 | 82.12 | 64.03 | 71.96 |
| 3    | 87.60 | 95.46 | 91.36 | 79.18 | 68.22 | 73.29 |
| 4    | 86.24 | 92.91 | 89.45 | 75.50 | 60.63 | 67.25 |
| 5    | 90.96 | 95.91 | 93.37 | 82.91 | 68.79 | 75.19 |
| 6    | 91.35 | 98.12 | 94.62 | 79.76 | 77.75 | 78.74 |
| 7    | 85.75 | 94.80 | 90.05 | 75.26 | 67.33 | 71.08 |
| 8    | 90.10 | 94.16 | 92.09 | 81.52 | 60.31 | 69.33 |
| 9    | 91.61 | 96.59 | 94.03 | 81.41 | 71.34 | 76.05 |
| 10   | 87.94 | 94.55 | 91.13 | 80.55 | 78.46 | 79.49 |

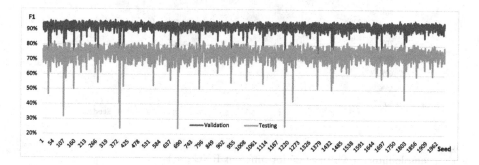

**Fig. 3.** Fluctuations of F1 scores achieved on validation and testing the test case with a sample interval width of 25 for different RNG seeds.

This was also confirmed for other RNG seeds, and the results show the behavior of local maximums and minimums presented in Fig. 2, as discussed in [15].

## 5.2   The Disparity Between Validation and Testing Results for Different RNG Seeds

Figure 4 presents more details on the achieved results for the first ten different RNG seeds. We notice a huge discrepancy between validation and testing results without any trend. The ranges between validation and testing F1 scores fluctuate between 11.64% and 22.83% in the example of the first ten RNG seeds.

Furthermore, we conclude that the validation results can not be used for optimization. For example, the maximal F1 score for validation is obtained for the RNG seed of two, while the maximum F1 score for testing is for the RNG seed of ten. In addition, the value of the F1 score for testing with the maximal validation F1 score is approximately close to the minimum, and the F1 score for validation with maximal testing F1 score is also close to the minimum of the validation F1 score. This peculiar behavior is observed for all analyzed sample interval widths and RNG seeds.

## 5.3   Model Stability for Different Features and Fixed RNG Seed

Figure 5 presents achieved results for different feature values (different width values in our case) and a fixed RNG seed of two and ten. Values of two and ten for RNG seeds have been chosen to present the model performance behavior for the RNG seed with a value of two with the maximal validation F1 score and minimal testing F1 score. An RNG seed with a value of ten achieves maximal testing F1 score and minimal validation F1 score.

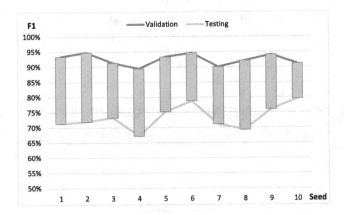

**Fig. 4.** Disparity of F1 scores between training and validation results on the test case with a sample interval width of 25 for different RNG seeds.

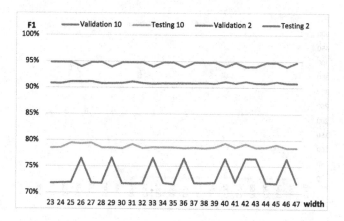

**Fig. 5.** Disparity of F1 scores between training and validation results on the test case with a sample interval width of 25 for different RNG seeds.

We observe relatively decreased performance fluctuations for validation during the training and increased fluctuations for testing performance. Interestingly, the absolute and relative range of F1 scores calculated as a ratio of the difference between the maximal and minimal F1 scores versus the average F1 score value is in the range between 0.07% and 3.08% for validation and between 0.20% and 6.79% for testing.

When the RNG seed is held constant, different feature set values do not affect the model performance significantly. This is somewhat unexpected, as it is commonly believed that distinct performance characteristics would result from training datasets. This may indicate that the model is relatively insensitive to variations in the training data, leading to an overly regularized or constrained model. Alternatively, it could suggest that the training datasets are similar in their distribution or the information they contain.

## 5.4 Validity of the Research Hypothesis

The results show that the models' performance depends heavily on the chosen RNG seed values. Moreover, the histogram of validation and testing F1 score values (Fig. 6) shows a distribution that looks like a Gaussian distribution, with a significant imbalance, skewed to the left from the peak.

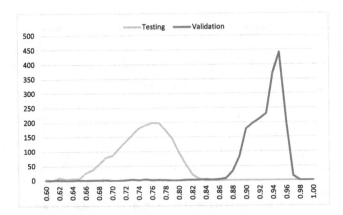

**Fig. 6.** Distribution of validation and testing F1 score values for different RNG seeds from 0 to 2000.

Evaluating the model for different RNG seeds from 0 to 2000, the achieved average F1 score for the validation dataset is 92.3%, and for the testing dataset is 73.80%. Table 5 shows the AVG and the reminder statistical parameters STD, RSD, MIN, MAX, RANGE, and RMSE, calculated versus the AVG.

We observe that the average values are not the distribution peaks. The relative standard deviation is up to 6.72% and the relative range of obtained values is 79.80% of the average value for testing F1 scores, which validates significant performance fluctuations. The RMSE of the validation F1 score is smaller than in the testing since the values are bounded by the maximum (reaching 100%). The main result of 24.61% RMSE for testing validates our hypothesis.

**Table 5.** Statistical analysis of obtained validation and testing F1 scores for different RNG seeds from 0 to 2000.

|  | AVG | STD | RSD | MIN | MAX | RANGE | RMSE |
|---|---|---|---|---|---|---|---|
| Validation | 92.33% | 3.23% | 3.50% | 58.05% | 96.71% | 41.87% | 10.42% |
| Testing | 73.80% | 4.96% | 6.72% | 23.79% | 82.68% | 79.80% | 24.61% |

This is significant because it suggests that the selection of RNG input can affect model performance and play a crucial role in achieving superior model performance. This could be especially useful in fields that require highly accurate models, such as medical diagnostics and financial forecasting. Furthermore, we conclude that the validation results in the training process do not lead to the best performance in testing.

Our hypothesis that RNG seeds using 1D Convolutional Neural Networks impact the performance results by more than 15% on the heartbeat classification performance (measured by RMSE) has been proven. The analysis within the research question shows that RNG seed substantially impacts the model performance since it affects the initialization of model weights, the shuffling of data for training and validation splits, and other stochastic processes within the model.

Adding a further hyperparameter will consequently increase the complexity of the model training process. Each additional hyperparameter adds a new dimension to the search space, thereby increasing the number of possible permutations and necessitating an increase in computational resources. This computational cost may involve additional training time, additional memory for storing the various models, and increased energy consumption. However, if this additional expenditure produces models with enhanced performance, it may be worthwhile. Despite the increased computational demand, the text's recommendation to experiment with the RNG seed as a hyperparameter could be a valuable optimization strategy for machine learning models.

## 6    Conclusion

We have realized an experiment of developing a 1D CNN DL model for the classification of V heartbeats with a feature set specified by width as several samples around the detected R peak. The test cases defined widths between 23 and 47. Each test case was executed with different RNG seeds, and the results were compared to validate the hypothesis that the RNG seeds impact the F1 score results by more than 15%. Furthermore, we investigate the level of impact of model performance for different RNG seeds and fixed width (feature). The evaluation of the testing dataset for the resulting F1 scores shows 24.61% RMSE.

We conclude that the RNG seed value plays a crucial role in model performance, balancing the need for apparent randomness in specific computational processes with reproducibility and consistency in experimental settings. The effect of the seed value on the results of particular algorithms emphasizes the need for caution when selecting and employing it. This seed value significantly impacts the model performance. The initial conditions and the inherent randomness of the training process affect the outcome. This emphasizes the importance of RNG seed selection and encourages the consideration of the seed as an optimizable hyperparameter.

Occasionally, the model's performance is more sensitive to its internal processes and initial conditions (as determined by the RNG seed) than to the specific data for the features it was trained on. This has significant implications for model validation and training methods. It emphasizes the need to manage and optimize the RNG seed during model training suggesting that the model may not be learning effectively from variations in the training data. In addition, our experiments confirm that optimization validating the training dataset does not optimize the testing performance. This may necessitate additional research to ensure that the model is adequately specified and that the training data is sufficiently diverse and representative.

This article does not aim to advocate for selecting feature sets from ECG samples or focusing on fine-tuning the RNG seed. The primary objective of this discussion has been to highlight the significant impact that different RNG seeds can have on the performance of ML models. We understand that the RNG seed needs to be recognized as an essential component of model training, deserving of thoughtful selection and careful management, opposite to an afterthought or a default setting. By drawing attention to this often-overlooked aspect, we aim to inspire a more nuanced and comprehensive approach to model training, resulting in enhanced performance and more reliable future results. Future work includes experiments letting the models train longer and stabilize to make a proper decision and highlight a potential double-dip phenomenon. We also plan to study different optimizers and network architectures which can further outline the influence of the initialization.

# References

1. Bock, S., Weiß, M.: A proof of local convergence for the Adam optimizer. In: 2019 International Joint Conference on Neural Networks (IJCNN), pp. 1–8. IEEE (2019)
2. De Chazal, P., O'Dwyer, M., Reilly, R.B.: Automatic classification of heartbeats using ECG morphology and heartbeat interval features. IEEE Trans. Biomed. Eng. **51**(7), 1196–1206 (2004)
3. Glorot, X., Bengio, Y.: Understanding the difficulty of training deep feedforward neural networks. In: Proceedings of the Thirteenth International Conference on Artificial Intelligence and Statistics, pp. 249–256. JMLR Workshop and Conference Proceedings (2010)
4. Hooper, S.M., et al.: Impact of upstream medical image processing on downstream performance of a head CT triage neural network. Radiol.: Artif. Intell. **3**(4), e200229 (2021)
5. Kiranyaz, S., Ince, T., Gabbouj, M.: Real-time patient-specific ECG classification by 1-d convolutional neural networks. IEEE Trans. Biomed. Eng. **63**(3), 664–675 (2015)
6. Li, F., Wu, J., Jia, M., Chen, Z., Pu, Y.: Automated heartbeat classification exploiting convolutional neural network with channel-wise attention. IEEE Access **7**, 122955–122963 (2019)
7. Li, Y., Liang, Y.: Learning overparameterized neural networks via stochastic gradient descent on structured data. In: Advances in Neural Information Processing Systems, vol. 31 (2018)
8. Liu, S., Papailiopoulos, D., Achlioptas, D.: Bad global minima exist and SGD can reach them. Adv. Neural. Inf. Process. Syst. **33**, 8543–8552 (2020)
9. Marrone, S., Olivieri, S., Piantadosi, G., Sansone, C.: Reproducibility of deep CNN for biomedical image processing across frameworks and architectures. In: 2019 27th European Signal Processing Conference (EUSIPCO), pp. 1–5. IEEE (2019)
10. Moody, G.B., Mark, R.G.: The impact of the MIT-BIH arrhythmia database. IEEE Eng. Med. Biol. Mag. **20**(3), 45–50 (2001)
11. Oxford Medical Education: ECG interpretation. https://oxfordmedicaleducation. com/ecgs/ecg-interpretation/
12. Ruby, U., Yendapalli, V.: Binary cross entropy with deep learning technique for image classification. Int. J. Adv. Trends Comput. Sci. Eng. **9**(10) (2020)
13. Sarvan, Ç., Özkurt, N.: ECG beat arrhythmia classification by using 1-d CNN in case of class imbalance. In: 2019 Medical Technologies Congress (TIPTEKNO), pp. 1–4. IEEE (2019)

14. Shi, H., Wang, H., Jin, Y., Zhao, L., Liu, C.: Automated heartbeat classification based on convolutional neural network with multiple kernel sizes. In: 2019 IEEE Fifth International Conference on Big Data Computing Service and Applications (BigDataService), pp. 311–315. IEEE (2019)
15. Silva, K.G., Aloise, D., Xavier-de Souza, S., Mladenovic, N.: Less is more: simplified Nelder-mead method for large unconstrained optimization. Yugoslav J. Oper. Res. **28**(2), 153–169 (2018)
16. Wang, Z., Yan, M., Chen, J., Liu, S., Zhang, D.: Deep learning library testing via effective model generation. In: Proceedings of the 28th ACM Joint Meeting on European Software Engineering Conference and Symposium on the Foundations of Software Engineering, pp. 788–799 (2020)
17. Xiaolin, L., Cardiff, B., John, D.: A 1d convolutional neural network for heartbeat classification from single lead ECG. In: 2020 27th IEEE International Conference on Electronics, Circuits and Systems (ICECS), pp. 1–2. IEEE (2020)
18. Xu, X., Liu, H.: ECG heartbeat classification using convolutional neural networks. IEEE Access **8**, 8614–8619 (2020)
19. Zhang, D., Chen, Y., Chen, Y., Ye, S., Cai, W., Chen, M.: An ECG heartbeat classification method based on deep convolutional neural network. J. Healthcare Eng. **2021**, 1–9 (2021)
20. Zhuang, D., Zhang, X., Song, S., Hooker, S.: Randomness in neural network training: characterizing the impact of tooling. Proc. Mach. Learn. Syst. **4**, 316–336 (2022)

# Dew Computing

# Disaster-Resilient Messaging Using Dew Computing

Anthony Harris(✉) ⓘ and Yingwei Wang ⓘ

University of Prince Edward Island, Charlottetown, PE C1A4P3, Canada
ajharris@upei.ca, ywang@upei.ca

**Abstract.** The speed of information transmission in our modern day and age requires us to stay connected more than ever. With seemingly endless forms of communication, our primary digital methods of email, telephony, and message services are not without vulnerabilities. Disaster can happen at any time, whether it be man-made or natural. Often, disaster results in the loss of electrical power, internet or telecommunication infrastructures. There currently exist several emergency communication protocols and many others in development but even they have their limitations and may not function under all circumstances. With this we propose incorporating the principles of dew computing to build a reliable, stable, and resilient mobile messaging application that increases the rate of successful transmissions. This paper explores dew computing's concepts of *Independence* and web service *Collaboration* by developing an application-level routing protocol we call *Spatial-Temporal Connection* (STC). By storing and relaying messages from both the primary mobile device and any neighboring devices, STC provides an alternate approach to staying connected when client-server network infrastructure breaks down.

**Keywords:** Dew Computing · Dew Server · Messaging Application · Ad Hoc Network · Disaster Communication · Spatial-Temporal Connection

## 1 Introduction

Disaster can take any shape or form. Pertaining to telecommunications, disaster results in disruptions ranging from a loss of power, downed cables, towers, or general electromagnetic interference. Hurricanes, earthquakes, tornadoes, and tsunamis are just a few of the natural disasters we as humanity face on a recurring basis. Disruptions can also be created by us. Overloaded networks, scheduled maintenance, or social-political issues can create the same scenarios. To remain prepared for such situations, we need to consider alternative communication methods.

Currently, there exists several disaster-resilient communication protocols in development incorporating mesh network architecture like mobile ad hoc networks (MANET) and small local area networks. These make use of the device's

Wi-Fi Direct, Bluetooth, or wired connection capabilities. These methods provide reliable connection and messaging services in most situations, however there are many instances when they are limited or even not possible.

Limitations to power, electronic interference, packet flooding, or even device type can mean not everyone has access to forming or joining these networks. There are also less drastic situations when these protocols are simply not necessary. Not all scenarios with limited to no connectivity should be classified as disasters. Travelling by land, sea, or air will undoubtedly cause periods with interrupted mobile service but they would not be categorized as disaster scenarios. It is in these cases dew computing can provide support for end users.

Beginning in 2015, dew computing emerged as scalable support system for internet and cloud-based networks by improving productivity in a distributed computing model [15]. Dew-cloud applications began to manifest as mobile computing evolved. Early forms of these applications brought cloud computing closer to the device level in the forms of offline data storage and web page access. Later examples of these applications include the file-sharing application Dropbox, the offline email storage for Gmail, and the income tax software Turbo Tax [10].

Coinciding with these applications, the focus of network optimization shifted toward more resilient and alternative client-server protocols. Some of these protocols developed into the form of Peer-to-Peer (P2P) networks - such as mesh networks, eventually elaborating to fully Mobile Ad hoc networks able to support the dynamicity of multiple nodes.

MANETs gained significant popularity in the following years - allowing users to create spontaneous, dynamic, and adaptable environments for communication [8]. These networks have remained generally restricted to military use and outlier functionality, but recently they have gained a mainstream following.

Recently, several messaging mobile applications began to incorporate the use of mesh architecture into their technology, including the now defunct *Firechat* [11] and *Bridgefy* [2]. As with many novel applications, several issues arose as these improvised networks reached certain user thresholds - dropped packets, loss of connection, and broadcast flooding. These common limitations exposed several vulnerabilities [9].

While optimizing MANET and mesh routing protocols in the form of Proactive, Reactive and Hybrid Routing creates more robust frameworks - vulnerabilities still exist, particularly when considering the plethora of mobile devices on the market and their varying technical limitations. This degree of unpredictability positions us to re-examine alternate messaging applications in environments of limited connectivity.

This paper provides a brief overview of the disaster-resilient technologies available today, their known vulnerabilities, and presents an alternate application incorporating the dew paradigm to challenge our understanding of communication routing. The mobile application presents itself as a messaging platform running our proposed application-level protocol STC based on the principles of dew computing's *Independence and collaboration* [14].

## 2   Network Vulnerabilities

In a standard Wide Area Network (WAN) model such as the internet, data travels instantaneously even if passing through multiple nodes or splitting into various paths along the way. Through wired or wireless connection, a user joins a network connecting them from their ISP to another. When this specific network is unavailable due to a disastrous scenario such as a loss of power, the remaining forms of network communication typically involve battery-powered mobile devices or UPS-powered servers and routers.

These devices can create small Personal Area Networks. A building running a back-up generator can continue to operate its own internal network of computers, routers, and servers. For mobile devices with capable Bluetooth, Wi-Fi or LoRa features, the possibilities include wireless personal area networks (WPAN) such as MANET or mesh networks.

Even though an improvised WPAN provides an excellent platform for personal devices to connect and exchange data, it does not automatically sync any new or modified information back to the internet when connection is restored. Additionally, issues like packet flooding, electromagnetic interference, battery life and processing power capabilities impact the chances for successful messaging.

### 2.1   Packet Flooding

As devices attempt to contact and create an improvised network, data packets begin to flood their transmission range. Even though modern routing architecture, like the hybrid *Zone Routing Protocol* can mitigate the risk of this by prioritizing its routing table search for local destinations, flooding still presents a serious issue as the network grows with every new device. The size of a WPAN can change drastically within seconds. Additionally, as Bluetooth radio range improves and extends, so will the number of potential relay nodes. The sheer volume of broadcast packets can disrupt essential communications resulting in dropped packets, channel blockages or denial of service [12].

### 2.2   Electromagnetic Interference (EMI)

Sometimes, even during regular activity, EMI disrupts radio signals from mobile devices. Other radio devices, LED/fluorescent lights, household appliances, lightning, solar flares, or radio jamming hardware can all impact and interrupt data transfer during times of disaster. WPAN are vulnerable to this phenomenon as they must continuously update and maintain their routing architecture. Rerouting or re-configuring the network can result in packet loss and network errors.

### 2.3   Battery Life

When a disaster strikes, the conditions can be unpredictable. Users rarely prepare with enough foresight. Mobile devices such as laptops, tablets and cell phones will have varying battery life at the time out an outage. If power remains unavailable, these devices may not possess enough battery life to incorporate themselves into a WPAN, especially if they are forced to enter a restricted power-saving mode. As a result, the battery continues to drain and the likelihood of a successful connection for message transmission diminishes over time [9].

## 2.4  Processing Power

One of the benefits of mesh networks is their decentralized architecture. Each device acts as a processor, receiver, and transmitter not only for its own messages but for other users as well. While this alleviates the issues present in a central-ized network, it comes at the cost of raw processing power required for ongoing maintenance. Mobile devices vary in terms of their individual processing power. Some laptops and smartphones can process vast quantities of data quickly. The trade-off for these devices is the demand on the system [1]. A device can crash if its processor is overwhelmed.

Accordingly, devices with limited processing functions such as a basic talk-and-text cell phone may intentionally be excluded entirely from the WPAN, even if it has Bluetooth or Wi-Fi capabilities. The proposed dew application seeks maximum inclusion from any type of mobile device with some capable form of radio transmitter.

## 3  Spatial-Temporal Connection

One major concern with improvised networks is the initialization. Not all devices have the capability to connect at a moment's notice due to limitations either inherit to their systems or created through signal interference. Even though they are adaptable and can reconfigure at will, most improvised networks require a direct transmission route to successfully transmit messages between users. If a user is offline or temporarily disconnected due to some inference event, the connection never exists. With this, we present a novel approach to messaging using dew computing.

Our application-level STC protocol sends and receives messages based on the mutable principles of *space* and *time* and comparable to delay-tolerant routing (DTN) protocols. With limited or no internet connectivity, messages transmit P2P via available wired or wireless connection including radio (Bluetooth, Wi-Fi, NFC and LoRa). If none of these methods are available, STC functions using the DTN *Store and forward* technique - the application's internal dew server physically holds and carries the message in a database table, routinely searching until a connection becomes available.

The transport protocol is essentially a novel protocol built upon several pre-existing transport protocols, mainly those associated with Bluetooth LE's *Generic Attribute Profile* and Wi-Fi Direct protocols (which will be later incor-porated in the development). Additionally, several adopted protocols can be configured based upon connection states of participating devices. For example, with full internet connectivity between devices, the application-level protocol STC would default transport to TCP/IP.

Initially, this application will be developed for Android OS using Java and Kotlin programming languages as these currently remain two of the most common languages in mobile application development. Upon further development, we plan to expand the application to support more mobile operating systems including iOS and KaiOS to allow for maximum exposure, testing and development [13]. The DBMS will be PostgreSQL - for its ability to utilize numerous data types like JSON and XML, its applicable add-ons such as PostGIS [3] to store geographic data for user location, and its expansive, open-sourced community [4].

The dew server's database consists of several tables; each one dictating a different action to take. The *pending table* holds the message if it is intended for a different user. The *received table* holds the message acknowledgment if it was successfully received. Finally, the *mailbox table* holds messages intended for its own user and displays them in the application's GUI.

As time is a mutable factor in the application-level protocol STC, this means communication consists principally of composed messages alongside geolocation and routing tables - live stream functions may not be supported in all scenarios with this DTN-like framework unless the devices connect directly over a mesh network with the permitting bandwidth. Another important consideration is the processing power. To accommodate as many device types as possible into the protocol, the data must factor in the lowest common denominator - both in terms of bandwidth capabilities and available device memory. These two factors will improve as the protocol is developed and more data and testing become available. Additionally, as more powerful devices enter the consumer market, the average user device will have increased capabilities. Presently, the application itself will utilize Bluetooth LE, as the lower energy requirements than standard Bluetooth will equate to a greater extension of battery life. JSON formatting will aid in keeping the transmissions light-weight and easily configurable.

While acting on both the behalf of the user and as an independent relay node, STC performs several of the functions of a MANET - such as message encryption and relay, but without the accompanying challenges of interference conditions, route discovery, processing power and constant battery power demands.

Using an identifier like the device's MAC address, the protocol tracks incoming and outgoing messages from all users, logging the time of activity, the GPS coordinates, and any other necessary information. This data exists in a database table processed by the dew server. Once connection is restored either through internet or another device sync, the table is updated. Over time, the application-level protocol STC transforms the device from a WPAN node into a WAN node. Suppose:

$$T = time \tag{1}$$

and

$$x = node \tag{2}$$

Then the time spent at each node($N$) would be:

$$N = T/x \qquad (3)$$

The route $AB$ can be calculated as,

$$\Sigma_{x=A}^{B} T = N_1 + N_2 + N_3 + ...N_n \qquad (4)$$

Figure 1 demonstrates user A broadcasting a message to user B using the proposed messaging application. In this example, internet and cellular service are both unavailable. In step 1, we have user A with a new message waiting to be sent to user B. The message is in the device's dew server database, in the *pending table*. Node 1 is within range of user A. User B is only within range of node 3. In step 2, user A sends a message to node 1. Its dew server identifies the message and places it in its own *pending table*. In step 3, node 1 moves closer to node 2. Once within range, node 1 forwards a copy to node 2. Like node 1, node 2 identifies the message has a different destination so it places it in its own *pending table*. Finally, in step 4, node 1 returns back to its original position. Node 2 moves within range of node 3 and relays a copy of the message. Node 3 repeats the processes of nodes 1 and 2. Since node 3 is already within range of user B, the message relays once more and arrives at its final destination. User B places the message in its *mailbox table* which is visible for the device user.

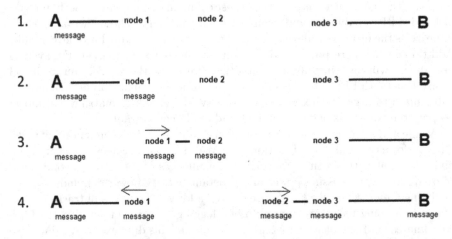

**Fig. 1.** Pathway of message from user A to user B.

Figure 2 demonstrates how a message acknowledgement is returned to the original sender over time with the aid of the intermediate nodes. We can still assume that internet and cellular service remain unavailable. In step 5, we see all five devices are still holding user A's message intended for user B. Since user B received the message, it broadcasts an acknowledgement for nearby devices. The instructions in this broadcast are simple - move the message from the *pending table* to the *received table* and relay this to the next device. If a node never held the original message, then this acknowledgement would be ignored. Node 3 is within range of both user B and node 2 so it receives the broadcast, moves the original message from *pending* to *received* and relays the broadcast to node 2 which does the same. In step 6, node 3 moves within range of node 1. As node 1 still has the message in its *pending table*, node 3 receives the original message a second time. Node 3's dew server knows that the message was already received by user B because of the *received table*. Node 3 broadcasts the acknowledgement to node 1. In step 7, node 1 moves the original message from *pending* to *received*. Since node 1 is also within range of user A, it relays this acknowledgement and completes the routing. Step 8 presents the completed acknowledgement routing from user B to user A.

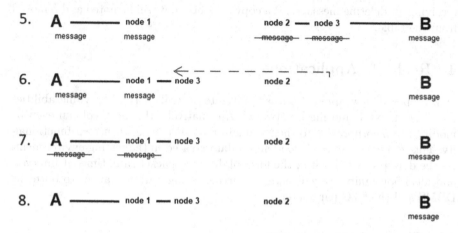

**Fig. 2.** Pathway of acknowledgement from user B to user A.

Even though message delivery is never guaranteed in a disaster scenario, the user's movement over time increases the probability of a successful device connection. Two devices entering Bluetooth LE proximity will exchange messages. Furthermore, to increase the likelihood of successful message receipt, we propose the protocol allows multiple devices to hold and relay the same message. This would alleviate concerns when one device loses power or is incapacitated by accident. Multiple backups would function in the same manner as storage-in-dew (SID) services and would be deleted upon reconnecting to the internet or receiving an acknowledgment receipt from the original sender.

Our proposed application will have flexible algorithmic parameters. As this transitions to an open-source project, we encourage developers to experiment with our dew server's protocol suite including message storage systems, number of relay attempts and time between searching for nearby devices. For instance, increasing the number of relay attempts might improve the number of successful messages received, however it might result in more packet flooding. Another example would be increasing the number of devices able to simultaneously hold the same message. If the ideal number falls too short, there may be a decrease in successful transmission rate. Conversely, too many devices may result in storage issues for low-end devices with lesser specifications.

A primary challenge with STC is the risk that many identical messages will co-exist across an indefinite number of devices. This form of message flooding can be mitigated by having an acknowledgement broadcast by the final recipient for a predetermined period, however like the previously mentioned challenge, there is no guarantee this will reach all impacted devices. One consideration is having each device create a countdown on each message it holds. This way, messages will self-delete over time. The user could have control over this feature with a delivery preset of $x$ *minute/hours/days* before deletion. The server would receive a "Message database" status from an intended recipient before sending, receiving, or deleting messages. If a copy is received, it will be noted and removed from the table.

## 4  Real-Life Applications

Our proposed dew application can mitigate several of the key vulnerabilities found in MANET and mesh networks. Alternatively, this protocol can even be modified to incorporate into their frameworks - increasing their own functionality. This will of course be dependent on data created in the testing environment. As the data presents itself in the form of dropped packet rate, timeout intervals and ideal hop limits, we will perform further configuration changes to both the DBMS and the STC parameters.

### 4.1  Disaster Communication

Under normal circumstances, it can be reasonable to assume cellular and internet services exist. When either or both are unavailable, this can be considered a disaster scenario. Figure 3 demonstrates such a scenario.

Suppose a devastating earthquake strikes a nation. Buildings collapse, people are injured, and all forms of internet and cellular telecommunications are severed. There is an immediate demand for emergency services but no method to call for aid. Almost every person could have a mobile device with Bluetooth and Wi-Fi capabilities with them, with any given amount of battery charge remaining. Even though a well-built MANET or mesh network may help those with the patience and power to build it and keep the network active, there could potentially be many users unable to connect for one reason or another.

Person X is trapped in a collapsed building. They are alone and sustained an injury. They have a mobile phone with some remaining battery life. They broadcast an SOS message which would include their geolocation. This message would have an arbitrary hop limit with the intention of reaching the most users possible without flooding the network. It may also have pre-set destinations it will attempt to reach like emergency contacts from the device itself.

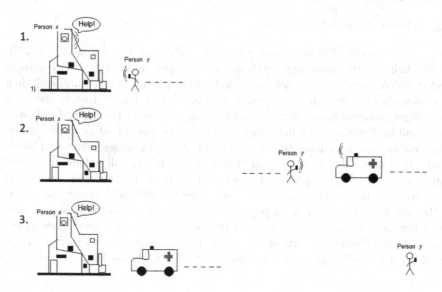

**Fig. 3.** 1) Person X requires emergency services. Person Y can accept Person X's broadcast and eventually 2) route it to an ambulance. 3) Ambulance arrives after viewing Person X's geolocation.

STC initializes immediately and searches for any neighboring nodes. Person Y is a bystander that was unharmed by the earthquake and now is actively searching for survivors. They are standing outside Person X's building and enter within Bluetooth range of their device. The application-level protocol STC of Person X's device may parse through the logic as follows:

1. An emergency broadcast: Is this message a broadcast? If yes, alert all devices within transmission range and display the broadcast message. If no, continue.
2. A contact: Is Person Y a contact of Person X? If yes, alert them and display the broadcast. If no, store the message in Person Y's STC database and hold until a new connection exists. Then, repeat this step for a pre-determined number of hops. Continue.
3. Continue or end: Whenever a broadcast acknowledgement is received, alert Person X. They can then choose to continue broadcasting, or to end the transmission and instead send termination acknowledgements back out to nearby devices.

In this scenario, Person Y sees the broadcast and can either respond via messaging, attempt to assist them and/or they can continue carrying the message until more help arrives. An ambulance is driving by at the time and enters within Bluetooth range of Person Y, but not Person X. The ambulance receives the rebroadcast from Person Y. The broadcast contains the location data of Person X and can now attempt to locate them. This is but one of many possible uses for the dew protocol.

## 4.2  General Travel

Beyond emergency situations, our proposed application has potential for additional, light-weight messaging services. For example, in the travel and tourism sector, the demand for internet access is becoming more of a requirement than a luxury. People now expect to remain connected with family, friends and even work through social media, e-mail, and texting. No matter the method of travel, users will undoubtedly at some point experience periods of interrupted service. Our dew application allows users to send and receive messages even when bandwidth is limited. Using cruise ship travel, users are typically charged for any internet usage, often at premium rates. In this scenario, the application-level protocol STC holds pending messages in its dew server until a sync can occur. If the user finds a connection or pays for only a few minutes of service, the messages sync automatically, saving costly time spent loading browsers, web pages, and login requirements. This is also done without the need to return to the application since the dew server is always running on the backend.

## 4.3  Snail E-mail

STC in many ways mirrors the principle of physical postal delivery, or *snail mail* as it is known colloquially. The following case makes use of the application-level protocol STC in the same way and can be refined to a similar colloquialism we propose as *snail e-mail*.

Suppose a remote village in a rural region has no telecommunication infrastructure, including no satellite internet service. The village may have two-way radios or ham radio, but no single villager has a method to send out an instant telecommunication at will. This village may in fact have residents who own cell phones or even temporary residents that visit periodically with digital devices, however there is no reception. Our application-level protocol STC could provide infrastructure for a physical message transport system to the villagers periodically. A bus or other vehicle travelling to or even near the village would carry the messages back to an area with service.

In Fig. 4, we can imagine a bus that visits a rural village in the middle of a country with no services. Here, the villagers can use their application to create and send messages, which are stored in the dew server of their device. Once the bus comes within range of their Bluetooth, the messages are relayed. The driver heads back into a town with internet or cellular service and the messages continue their routing until they reach their destinations. The same effect happens in reverse.

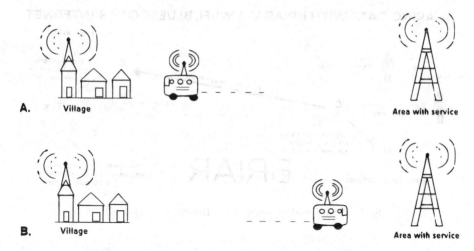

**Fig. 4.** Pathway of message from Village to an area of service and back again.

## 5   A Case Study - Briar

The Briar Project [5] took off from where Firechat ended. Expanding upon the mesh framework, Briar includes APIs allowing mobile devices to sync over The Onion Router (TOR) network - with an internet connection - for even more secure communication. The P2P, encrypted messaging and forum application is built upon the following principles:

– The messages are stored in the device itself. They are not stored in the cloud
– Devices connect directly with each other without an intermediary such as an ISP
– The software is free to use and is openly available to the public

Briar was originally intended for users requiring a secure, decentralized form of messaging, especially in scenarios with strong media censorship. Many servers and telecommunications are controlled by corporations or governments, consequently it is imperative that messaging remains secure and free from eavesdropping.

Briar uses the Bramble protocol suite including Bramble Handshake Protocol, QR Code Protocol, Rendezvous Protocol, Synchronization Protocol, and Transport Protocol that is specifically designed for delay-tolerant networks [7].

While no method of digital communication is completely sheltered from exposure to hackers, Briar allows trust between its users as the software is free and open-sourced.

Figure 5 displaying Briar's routing mechanism shares many of the same features in our proposed protocol including holding, relaying and physically transporting messages from one device to another.

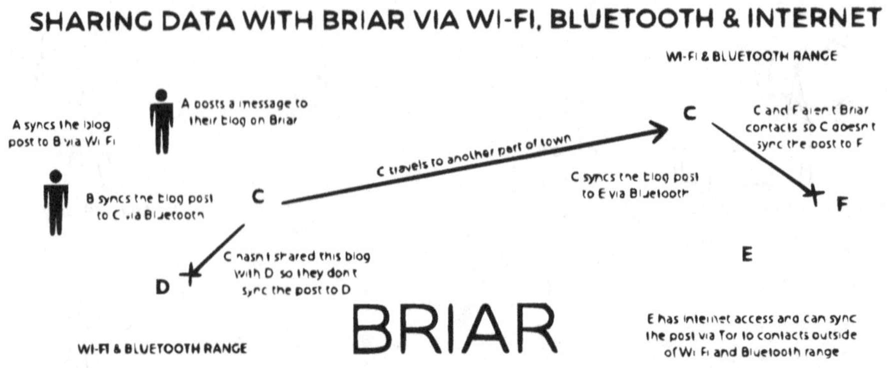

**SHARING DATA WITH BRIAR VIA WI-FI, BLUETOOTH & INTERNET**

**Fig. 5.** Pathway of message using Briar application [5].

When we apply Briar to our examples of disaster communication, general travel, and snail e-mail, it performs nearly identical to our proposed dew application. Briar's routing structure permits emergency communications and messages can be stored in the form of blog posts when connections are unavailable. Unfortunately, Briar falls short in three key areas:

### 5.1   Contacts

Briar allows three messaging formats: Private Group, Forum, and Blog. To send a message to another device, Briar requires that both users must meet the following criteria [7]:

- They must be a contact
- They must also subscribe to the group/forum/blog where the message was posted
- They must also have accepted the other contact as a subscriber via an invitation.

Undoubtedly, this presents privacy concerns among devices where trust cannot be implied before sending or receiving messages. Unless the destined party is in direct range for P2P communication, the sender will need to add every unknown advertising device as a contact to increase the likelihood of their message successfully reaching its target. This creates issues during emergencies when parties do not have time to add new contacts. This can be resolved with our proposed dew application. Instead of manually adding every potential relay device as a contact, the application functions as a trustless system, providing message relay without user knowledge or approval - as the dew server will handle the relaying on the backend.

### 5.2   Destination

As previously mentioned, posts are seen by all group/forum/blog contacts when synced up. If a user creates a private group to send a direct message, transmission is only possible over TOR or with a direct Bluetooth/Wi-Fi connection. Ideally, users should be able to send and relay messages with or without internet access and without the need to have pre-existing contacts between nodes. If the application works as a trustless system, messages can be encrypted and sent to other devices without other users having the ability to read them.

### 5.3   Offline Syncing

With no internet connection, messaging a Briar contact outside of the network range requires all users to be contacts to the sender. This would hypothetically allow message holding and relaying via contact syncing. Unfortunately, we learned this is very inefficient and complicated from the previous sections. Fortunately, Briar is developing a *Briar Mailbox* [6] component which will act in a similar way to the Gmail application. It functions on any PC or Android device using the principles of dew computing. Once fully developed, it may be able to resolve many of the previously discussed issues. However, it should be noted that Briar Mailbox still requires an internet connection to sync messages between devices. To allow mailbox syncing even while offline, this application would need to be combined with Briar messenger itself and not act as a separate entity. Our dew application will allow user mailboxes to sync to each other over Bluetooth or Wi-Fi when in proximity, regardless of contact status. If a device detects another mailbox containing a pending message, it will copy and send it.

## 6   Conclusion

This paper examined several current solutions to disaster scenario communication and proposed an alternate approach using dew computing. In uncovering the inherent vulnerabilities found within existing MANET and mesh network architectures, our application-level protocol STC addresses several of these concerns, however its potential can also be found in other applications. From periods of limited connection during general travel, to a simplified method of communication via *snail e-mail*, STC and dew computing can be seamlessly integrated into many modern technologies.

As this project emerges in the open-source community, we will encourage interested developers to explore additional applications of its use. It is currently being developed for Android, in the languages of Java and Kotlin, but we plan to extend its functionality to all mobile operating systems if possible.

Upon our messaging application's open-source release, we will encourage the community's input in areas of optimization including:

– Internal storage systems and database structures
– Advertise, Connect and Scan settings
– Optimal number of relay attempts
– Ideal hop limit
– Message storage file size limit, and

– Message storage time limit

This list is not exhaustive in any sense. As we continue to explore this application of dew computing - and with subsequent novel technologies which coincide with it - more of its features will be explored.

STC configures in every sense the dew concepts of *Independence* and *Collaboration*. This application is capable of transmitting messages in a decentralized framework when network connections are down. When connectivity is restored, it syncs and updates its database - relaying any pending messages via TCP/IP.

While the onus of addressing all the vulnerabilities of mesh and ad hoc networks may not yet fall entirely on the proposed application, we instead present it as an alternative approach. Additionally, the STC protocol's performance in other messaging scenarios with limited connectivity such as *general travel* and *snail e-mail* demonstrates its value and potential. With further development, we expect decentralized and disaster-resilient messaging services to continue to propagate and gain ubiquity in our increasingly-connected and technologically evolving society.

# References

1. Akyildiz, I., Wang, X.: A survey on wireless mesh networks. IEEE Commun. Mag. **43**(9), S23–S30 (2005). https://doi.org/10.1109/MCOM.2005.1509968
2. Hagenhoff, K., Viehmann, E., Rodosek, G.D.: Time-sensitive multi-flow routing in highly utilized MANETs. In: 2022 18th International Conference on Network and Service Management (CNSM), pp. 82–90 (2022). https://doi.org/10.23919/CNSM55787.2022.9964689
3. PostGIS: About postgis. https://postgis.net/. Accessed 30 July 2023
4. PostgreSQL: About postgresql. https://www.postgresql.org/about/. Accessed 30 July 2023
5. Project, B.: How it works (2018). https://briarproject.org/how-it-works/. Accessed 19 Apr 2023
6. Project, B.: Briar mailbox (2020). https://code.briarproject.org/briar/briar-mailbox/. Accessed 23 Apr 2023
7. Project, B.: Faq (2020). https://code.briarproject.org/briar/briar/-/wikis/home. Accessed 23 Apr 2023
8. Ray, P.P.: An introduction to dew computing: definition, concept and implications. IEEE Access **6**, 723–737 (2017)
9. Raza, N., Umar Aftab, M., Qasim Akbar, M., Ashraf, O., Irfan, M.: Mobile ad-hoc networks applications and its challenges. Commun. Netw. **8**(03), 131–136 (2016)
10. Rindos, A., Wang, Y.: Dew computing: the complementary piece of cloud computing. In: 2016 IEEE International Conferences on Big Data and Cloud Computing (BDCloud), Social Computing and Networking (SocialCom), Sustainable Computing and Communications (SustainCom) (BDCloud-SocialCom-SustainCom), pp. 15–20 (2016). https://doi.org/10.1109/BDCloud-SocialCom-SustainCom.2016.14
11. Soares, E., Brandão, P., Prior, R., Aguiar, A.: Experimentation with MANEts of smartphones. In: 2017 Wireless Days, pp. 155–158 (2017). https://doi.org/10.1109/WD.2017.7918133

12. Soomro, A.M., et al.: Comparative review of routing protocols in manet for future research in disaster management. J. Commun. **17**(9) (2022)
13. Statcounter: Mobile operating system market share worldwide. https://gs.statcounter.com/os-market-share/mobile/worldwide. Accessed 30 July 2023
14. Wang, Y.: Definition and categorization of dew computing. Open J. Cloud Comput. (OJCC) **3**(1), 1–7 (2016)
15. Wang, Y., Skala, K., Rindos, A., Gusev, M., Yang, S., Pan, Y.: Dew computing and transition of internet computing paradigms. ZTE Commun. **15**(4), 30–37 (2017)

# A Review of Dew and Edge Computing: Two Sides of a Modern Internet of Things Solution

Marjan Gusev$^{(\boxtimes)}$ (ID)

Faculty of Computer Science and Engineering, Sts Cyril and Methodius University in Skopje, Skopje, North Macedonia
marjan.gushev@finki.ukim.mk

**Abstract.** Dew and edge computing are the sophisticated post-cloud architectural approaches that bring computing closer to the user for applications addressing the Internet of Things. In this paper, we analyze the requirements of post-cloud architectures to build such a solution, which clarify the main differences between dew and edge computing approaches. The analysis includes energy consumption, communication and processing requirements, latency, and throughput, and the evaluation shows how these requirements impact performance. In addition, we also analyze architectural approaches, including hardware/software coexistence, scalability, hardwareless computing, virtualization, interoperability, and portability. This research will check the validity of a hypothesis whether the dew and edge computing two sides of the same modern Internet of Things solution.

**Keywords:** Dew computing · Edge computing · Fog computing · Cloudlet · Mobile Edge Computing

## 1 Introduction

Post-cloud computing refers to the change of the location where the computing requirements will be fulfilled. Mainly, the idea is to change the mindset and use concepts of distributed systems nearby the user instead of centralized data centers [27]. A lot of different solutions have been developed for the Internet of Things (IoT) applications [7], including cloudlets locating smaller nearby servers by Internet providers [28], fog computing addressing the communication infrastructure to set a smaller nearby server by a telecom operator [3], edge computing as a general approach to set smaller nearby servers at the edge of the network [29], mobile edge computing locating smaller nearby servers at base stations by mobile operators [18], dew computing as a general term for computing even out of the Internet network [36]. A simplified explanation is that edge means bringing the computing closer to the user, with specific implementations of cloudlets (by Internet providers), fog computing (by network telecom providers), or mobile edge computing (by mobile operators), whereas dew computing to be considered

M. Mihova and M. Jovanov (Eds.): ICT Innovations 2023, CCIS 1991, pp. 94–107, 2024.
https://doi.org/10.1007/978-3-031-54321-0_7

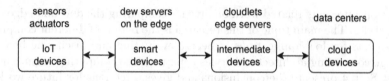

**Fig. 1.** A simplified 4-layer architecture of post-cloud solutions for IoT.

as an upgrade of the edge computing concept with a possibility of autonomous performance.

A typical post-cloud system architecture usually consists of the following layers (Fig. 1):

- *IoT devices* operated by end users,
- *Smart devices* on the edge of the Internet,
- *Intermediate computing devices* enabled by Internet providers, and telecom or mobile operators to serve closer to the edge of the network, and
- *Cloud data centers* for complex processing and permanent storage,

Several implementations can be identified:

- *IoT standalone* solution when the user operates the IoT device by a nearby smart device,
- *Edge computing solutions* that include cloudlets, fog and mobile edge computing, and
- a *Cloud-operated IoT solution* that processes all transferred data.

In this aspect, we refer to dew computing as a specific architecture that canserve as IoT standalone (autonomous solution) and an edge computing solution. Furthermore, we set the focus on analyzing requirements for edge and dew computing solutions.

In our earlier paper [11], we discussed the difference between dew and edge computing and why dew computing is a broader architectural concept compared to edge computing, covering applications for IoT devices out of the Internet network perimeter. In this paper, we continue analyzing the features that make the fundamental differences between edge and dew approaches, focusing on the computing requirements for IoT applications. Particularly, we address

1. networking, computing, storage, and energy consumption, as *resource requirements*;
2. throughput, response time (latency) as *performance-related requirements*;
3. autonomy, hardware/software coexistence, scalability, and hardwareless computing as an *architectural approach*, and analyzing the
4. virtualization, interoperability, and portability as *essential properties* of such systems.

We set a research hypothesis to check if the dew and edge computing are two sides of the same solution. The comparative analysis in this review of requirements for implementing an IoT solution will check the hypothesis validity.

Section 2 presents related work on articles comparing the dew and edge computing styles. The main topic of this research is the review of different computing requirements for IoT applications in Sect. 3 evaluating and discussing how dew and edge computing architectural styles cope with the computing requirements. Finally, Sect. 4 presents the conclusions and gives directions for future work.

## 2   Related Work

Several researchers analyzed post-cloud architectures, specifying the definition and main building concepts of cloudlet, dew, edge, and fog (CDEF) computing [37]. explaining the differences between the origin, definition, principles, and application domain [21], leveraging the capabilities of heterogeneous devices and edge equipment, including network bottlenecks, and demand of new complex application scenarios [39].

Studies that explain the whole computing architecture space include: moving from the grid and cloud computing through the fog and edge computing towards dew computing [1], the complementary piece of cloud computing [25], a hybrid adaptive evolutionary approach to distributed resource management in dew based edge to cloud computing ecosystem [26], the rainbow: integrating cloud, edge, fog, and dew computing into the global ecosystem [33].

An introduction to dew computing has been analyzed from the perspective of the main concepts and implications [22], including the impact, potentials, and challenges for power management, processor utility, data storage, operating system viability, network model, communication protocol, and programming principles. Dew architecture is based on the concepts of a dew server with independence from the upper layer servers collaborating with them when the Internet is enabled, concepts well defined by [31,35].

The concepts of IoT devices and applications are mixed with the architectural design of post-cloud computing. DEW is introduced as a new edge computing component for distributed dynamic networks. The system architecture consists of the application layer (cloud services), network layer (fog, edge), and perception layer (mist computing, dew computing, and edge devices) [4]. This is different from architectural approaches in other literature and especially from dew computing concepts elaborated by another research [8,25,31,35,36].

Applications on IoT devices as an evolution of edge computing have been analyzed by the location of the processing resource, computation and storage offloading, energy consumption, communication, and processing requirements [14]. The continuum from cloud to IoT devices (things) was analyzed [17] with the opportunities and challenges in cloud, fog, edge, and dew computing.

A focus on differences between edge and dew computing has been analyzed in our earlier paper [11] with an analysis of the location of the processing resource, synchronization, and offloading concepts from a computing perspective, energy consumption, processing power, and storage capacity. In addition, the analysis includes other related features, such as data and computation replication and caching based on space and time locality. The main differences have been in the availability of an application to be independent on the cloud server and the possibility for stand-alone and autonomous performance.

Applications of dew and edge computing were analyzed from different perspectives, such as offline computing architecture for healthcare IoT [19], survey, e-healthcare case study and future direction of edge computing for IoT [23], or developing an AI doctor dew computing application at the edge [12], or for the Medical Internet of Things [16]. Other application domains include dew computing for advanced living environments [32], dew computing architecture for cyber-physical systems, and IoT [9].

Edge computing has been redefined by looking at the edge based on dominant data flows, which reduce entropy across locations and ensure overall system health [24]. However, [5] concludes that dew computing is an added value as a new approach to cloud-dew architecture.

Fog computing was analyzed from architecture to edge computing and big data processing [30], a combined edge/fog computing implementing an IoT Infrastructure [15]. The fog computing paradigms and scenarios have been analyzed with a focus on security issues [34]. A systematic review of security and privacy requirements in edge computing [38] presents state-of-the-art and future research opportunities for all post-cloud conceptual architecture approaches, which include edge, fog, mobile edge, cloudlet, mist, and dew computing.

## 3   Analysis of Architectural Features

We analyze how different technologies impact post-cloud architectures, including networking, computing, energy consumption, throughput and response time, autonomous performance, hardware/software coexistence, scalability, hardwareless computing, and underlying system properties, such as virtualization, interoperability, and portability.

### 3.1   Architectural Approach

This research refers to a classical 4-layer architecture of post-cloud IoT solutions (Fig. 1) based on the location of processing resources comprising of 1) IoT devices (dew devices), 2) Smart devices (edge devices and dew servers), 3) Intermediate cloudlets and edge servers, and 4) Cloud data centers.

Furthermore, we analyze several solutions identified:

- *IoT standalone*, a solution that uses only the lowest two architectural layers (Fig. 1),
- *Cloudlet*, a 4-layer solution that locates the processing on resources available by Internet providers and uses the cloud for the result (data) sharing,
- *Fog/Mobile edge solution*, a 4-layer solution that locates the processing on resources available by telecom or mobile operators and uses the cloud for the result (data) sharing,
- *Cloud-operated IoT*, a complete 4-layer architecture.

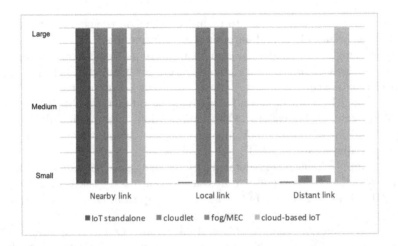

**Fig. 2.** Communication requirements for post-cloud solutions of IoT.

Furthermore, according to this specification, the edge computing solution is realized by the whole 4-layer architecture (Fig. 1) and dew computing as a combination of a 2-layer architecture (IoT standalone) as an autonomous device and a 4-layer architecture in case of Internet availability,

## 3.2   Networking Perspective

Intermediate devices were initiated by the requirement to use a small IoT device with a limited power supply and, therefore, small computing and networking capabilities to keep low energy consumption. Figure 2 gives an overview of communication requirements for different implementations of post-cloud IoT solutions.

Values in Fig. 2 represent only the requirement expressed in how much data needs to be transferred without analysis of the speed and acceptable delay. Of course, everybody would expect the desired transfer speed to be at the highest possible level the technology can offer. The reality is that it will consume a lot of energy.

Available technology for radio communication is analyzed in Table 1. The range is the feature that dictates the energy consumption of the attached device.

Personal area networking (PAN) is used between the lowest two layers of the 4-layer post-cloud system architecture. Bluetooth or direct cable connections are preferable to PAN technology to realize low-energy communications and save the battery. We consider only low-energy Bluetooth technology, omitting class 1 devices. ZigBee is used for home automation with throughput up to 250 Kbits/s, while Bluetooth reaches 1 Mbits/s or Bluetooth 3 up to 24 Mbits/s. However, Bluetooth does not provide high speeds that other radio communications can,

**Table 1.** Performance analysis of communication technologies for post-cloud solutions of IoT.

|  | Range | Energy | Speed | Latency | Delay |
|---|---|---|---|---|---|
| Bluetooth | nearby (<10 m) | low | small | low | low |
| WiFi | local (<50 m) | high | high | low | low |
| 3G/4G | medium (<30 km) | high | high | low | low |
| WAN | distant | high | high | high | high |

including WiFI 802.11n up to 150 Mbits/sec or 802.11ac up to 860 Mbits/s with low latency.

Local area networking (LAN) is used between smart and intermediate devices, such as cloudlets or edge servers. LAN technology achieves the highest speed, mainly due to the small distances, compared to Wide area networking (WAN), due to the distance to the cloud server. The latest radio area networking (RAN) technology offers high throughput in LANs, and it is expected to reach even higher values with 5G. Currently widely available are WiMax 2, with a throughput of 183 Mbits/sec, and 4G LTE 12 of 1 Gbits/sec.

WAN technologies are mainly used for long-distance communication between intermediate devices and cloud data centers, usually by high-speed optical cables.

Table 1 summarises the performance analysis of used technologies and explains the choice of the corresponding technology.

### 3.3   Computing Requirements

Processing power and storage are dominant features analyzed as computing requirements, but not the only ones. There are also other hardware and software parts, such as the operating system for user interface, power management, internal control of background and foreground functions, authentication and authorization, etc. We specify a scenario about the distribution of computing requirements:

- signal and data processing - large,
- storage - small,
- authentication, authorization, and internal control - small, and
- operating system (user interface, power management) - small.

Figure 3 gives an overview of processing requirements for different implementations of post-cloud IoT solutions. An IoT device performs several functions, and in this paper, we analyze it as a device for sensing, controlling mechanical or other parts, processing the signal, and communicating with devices at the upper architectural level to exchange data. In the case of an IoT standalone solution, the IoT device will perform almost all computing requirements. In other analyzed implementations, it will just offload all data to the devices in upper layers, so the storage and data processing will be performed elsewhere.

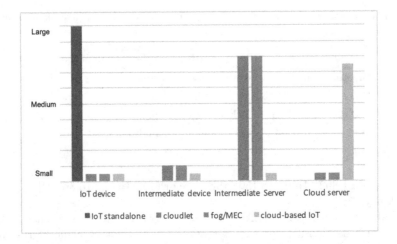

**Fig. 3.** Processing requirements for post-cloud solutions of IoT.

The intermediate device and server perform data processing in edge computing scenarios and only share the results with the cloud data center. The data center performs data processing in the cloud-based implementation, and the intermediate devices only offload data and processing requirements to upper levels (Fig. 3).

### 3.4 Energy Consumption

In this analysis, we consider only the energy consumption due to the processing and communication requirements, without reference to the sensing or actuator controlling functions, since they will be present in all implementations.

The following needs to be analyzed for energy consumption. Denote by $p_{IoT}$ the coefficient of the amount of computation to be processed on the IoT device and by $c_{IoT}$ the amount of data to be transferred to the intermediate device. Assume that $N_{comp}$ is the total number of computations and $N_{data}$ is the total number of data to be transferred to the intermediate device.

Total energy $E_{IoT}$ consumed on the IoT device can be calculated by (1), where $E_{proc}$ and $E_{comm}$ are the energy consumption to process a single computation or transfer a single data item.

$$E_{IoT} = p_{IoT}N_{comp}E_{proc} + c_{IoT}N_{data}E_{comm} \tag{1}$$

Note that the compromise to save energy on the IoT and choose what to offload will depend on the selection of $p_{IoT}$ and $c_{IoT}$. One needs to find their optimum value by the values of $E_{proc}$ and $E_{comm}$. Although the producer of the IoT device generally gives data about energy consumption, most applications are experimentally verified to find the optimal value.

Finally, the storage capacity is another factor in the comprehensive optimization approach to determine what to offload. Increasing the storage capacity consumes more energy.

### 3.5 Throughput and Response Time

Energy consumption is not the only parameter to be analyzed to determine what to offload and distribute the processing on a specific device. Performance analysis of the speed requirements needs to be considered since the processing power of the small IoT device is much lower than the one found on the intermediate device or the servers. Denote by $T_{comp}$ the time required for a single computation (assuming that all calculations take the same time) and by $T_{comm}$ the time needed to transfer a single data element to the intermediate device.

The total time to process all data on the IoT device is expressed by (2).

$$T_{IoT} = p_{IoT} N_{comp} T_{proc} + c_{IoT} N_{data} T_{comm} \qquad (2)$$

Assuming that the intermediate device will perform the remaining computations (without replication) in time $T_{id}$ then the total time is expressed by (3). When the intermediate device transfers all data to a more distant server or cloud server, this can be added to the analysis, but for simplicity, assume that all the computation will be performed locally as in post-cloud solutions.

$$T_{total} = T_{IoT} + T_{id} \qquad (3)$$

Finally, the optimization will require finding a minimum of $T_{total}$ and $E_{IoT}$ by determination of the coefficients $p_{IoT}$ and $c_{IoT}$. This solves the problem of what and where to offload, especially for a comprehensive analysis assuming the possibility to offload to the intermediate or cloud server.

### 3.6 Hardware/Software Coexistence

The question of which functions will be integrated into the hardware (IoT device) and which will be realized as software for the computing requirements determines the hardware/software coexistence and the proportion of what will be built as hardware or software. In the four analyzed post-cloud implementations, the decisions will be made by analyzing the computing requirements to make a trade-off towards better performance, more functionalities, and lower energy consumption.

The target of an IoT standalone solution is to provide autonomous performance with or without an Internet connection, which is implementing a dew computing concept. In this case, more functions will be built into hardware or on the IoT software. However, the edge computing solutions with edge servers provided by Internet providers (cloudlets) or mobile operators (fog or mobile edge computing) will integrate fewer functions on the IoT device, with minimal software mainly for internal control and transferring data to the neighboring devices.

Deciding which part of the software will be built by hardware (or the internal IoT device operating system) depends on power management and energy consumption. The designers of a battery-operated device will build less software in the IoT device (hardware). In contrast, the IoT device is connected to the energy grid, then this might integrate more powerful functions with complex processing.

### 3.7 Scalability

In this analysis, we refer to scalability as the ability of a system to cope with increased/decreased workload, mainly realized by adding additional computing resources. Scalable dew systems were analyzed in our earlier paper [13], with an example in telemedicine to build a system that copes with thousands of incoming streams.

Scalable distributed computing hierarchy was analyzed by specifying the differences between cloud, fog, and dew computing [31]. Furthermore, we analyze only implementations of scalable dew and edge computing approaches, considering that a scalable system can process large workloads. We add a requirement to specify the target to keep the same performance by selecting from many approaches. In our case, the target of an IoT system will be to keep the same throughput or response time (delay).

The scalability in the IoT standalone implementation can be observed by the smart device being capable of processing more IoT sensors or engaging similar nearby devices and offloading an excess of computing requirements (horizontal scalability).

The scalability of both edge computing solutions is mainly realized in a vertical direction, offloading computing requirements to the edge servers on the perimeter of the Internet network. Note that edge computing currently does not support horizontal scalability, viable to be realized soon by offloading computing requirements to a nearby device on the same architectural layer, essential for the dew computing solution.

### 3.8 Hardwareless Computing

The idea of not worrying about resources and building a serverless architecture [2], has been implemented on an edge computing platform for IoT [20]. The concept of hardwareless computing was introduced [10] as a generalization of the inftrastructureless computing approach, which is different from the "infrastructureless communication" used for device-to-device communications [6]. Analyzing the communications, two devices can communicate without the Internet infrastructure. In contrast, in the context of computing, the user can perform the computing requirements without worrying about the infrastructure or not being aware of the computing realized on a nearby device, server, or distant data center.

Focusing on the scope of edge and dew computing, we refer that the computing requirements are fulfilled closer to the source on the edge of the network (edge computing) or on the smart device itself (dew computing concept of the

IoT standalone without the presence of the Internet network). To distinguish between these two concepts used in the context of edge and dew computing, we provide a rather simplified practical user-centric explanation:

- *Serverless computing* - I donÕt care how many edge servers or distant cloud data centers will be used since I know the requirements will be processed as fast as possible.
- *Deviceless computing* - The smart device connected to the IoT device can not deliver the expected performance and searches for a nearby device to take over fulfillment of the computing requirements.
- *Thingless computing* - The IoT device (thing) is aware can not deliver the expected performance, and the embedded operating system searches for a nearby host (IoT device) to take over the fulfillment of the computing requirements.

Hardwareless computing is tightly connected to scalability as a system property that can cope with increased workflow. However, it can also provide a system property to cope with cases of limited power supply, which is more relevant for dew computing than edge computing.

### 3.9 Underlying Technology and System Properties

We specify other features needed to implement scalability and hardwareless computing concepts and fulfill the computing requirements on edge and dew computing solutions for IoT:

- *Virtualization* is the essential technology for implementation, including server virtualization for serverless computing, device virtualization for computing, and on the lowest scale, virtualization on the level of things (IoTdevices). Although server virtualization is a mature technology, the virtualization of devices and things are hot research topic due to the involvement of heterogeneous resources.
- *Interoperability* is another property specifying standards for developing and transferring data and code recognized and implemented by different providers. This leads to an open platform where applications can use or amalgamate services and infrastructure from another system.
- *Portability* is a system property for efficiently migrating virtual instances on different levels from one host to another. This is extremely important for implementing hardwareless computing with standards to exchange hosts or engage more hosts.

## 4   Conclusion

A simplified view of a modern IoT *post-cloud system* is presented in Fig. 4 specifiying architectural and technology concepts and identifying the following:

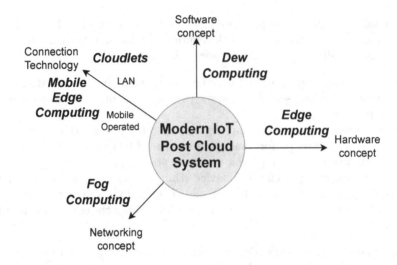

**Fig. 4.** Architecture Concepts and Technology of a Modern Post-Cloud IoT system.

- *Edge computing* is a hardware architecture approach,
- *Fog computing* is a networking concept,
- *Mobile edge computing* is a smartphone concept,
- *Cloudlet computing* is a LAN concept, and
- *Dew computing* is a software architecture concept applied to edge computing.

Analyzing the edge and dew computing implementations of IoT applications, we conclude that dew computing adds a new property of an edge computing solution to fulfill the computing requirements even in a 2- layer architecture of an IoT standalone system or perform in a classical 4-layer architecture of a post-cloud system where the smart devices are on the edge of the Internet and processing is realized on intermediate devices and servers.

Both approaches are analyzed for communication, processing, and energy consumption to deliver expected throughput and response time. In this case, we conclude that autonomous performance in dew computing is an added value to the edge computing implementation.

Analyzing the scalability and hardwareless computing concepts, we conclude that the underlying virtualization technology and system properties of interoperability and portability coping with standards are essential for developing horizontal and vertical scalability and "take over" functions. In this context, server virtualization is a reality, while smart device and thing (IoT device) virtualization is at the dawn of its evolution.

We presented solutions and open issues for achieving scalable hardwareless solutions realized by edge and dew computing implementations.

Finally, we confirm the research hypothesis that both dew and edge computing, respectively as software and hardware approaches, are just two sides of a modern IoT solution.

**Acknowledgment.** The experiment "CardioHPC - Improving DL-based Arrhythmia Classification Algorithm and Simulation of Real-Time Heart Monitoring of Thousands of Patients" has received funding from the European High-Performance Computing Joint Undertaking (JU) through the FF4EuroHPC project under grant agreement No 951745. The JU receives support from the European Union's Horizon 2020 research and innovation programme and Germany, Italy, Slovenia, France, and Spain.

# References

1. Ageed, Z.S., Zeebaree, S.R., Sadeeq, M.A., Ibrahim, R.K., Shukur, H.M., Alkhayyat, A.: Comprehensive study of moving from grid and cloud computing through fog and edge computing towards dew computing. In: 2021 4th International Iraqi Conference on Engineering Technology and their Applications (IICETA), pp. 68–74. IEEE (2021)

2. Baldini, I., et al.: Serverless computing: current trends and open problems. In: Chaudhary, S., Somani, G., Buyya, R. (eds.) Research Advances in Cloud Computing, pp. 1–20. Springer, Singapore (2017). https://doi.org/10.1007/978-981-10-5026-8_1

3. Bonomi, F., Milito, R., Zhu, J., Addepalli, S.: Fog computing and its role in the internet of things. In: Proceedings of the First Edition of the MCC Workshop on Mobile Cloud Computing, pp. 13–16. ACM (2012)

4. Cristescu, G., Dobrescu, R., Chenaru, O., Florea, G.: DEW: a new edge computing component for distributed dynamic networks. In: 2019 22nd International Conference on Control Systems and Computer Science (CSCS), pp. 547–551. IEEE (2019)

5. Fisher, D.E., Yang, S.: Doing more with the dew: a new approach to cloud-dew architecture. Open J. Cloud Comput. (OJCC) **3**(1), 8–19 (2016)

6. Fodor, G., et al.: Design aspects of network assisted device-to-device communications. IEEE Commun. Mag. **50**(3), 170–177 (2012)

7. Gusev, M.: Edge and dew computing for streaming IoT. In: Proceedings of The 3rd International Workshop on Dew Computing, pp. 1–7 (2018)

8. Gusev, M.: Formal description of dew computing. In: Proceedings of The 3rd International Workshop on Dew Computing, pp. 8–13 (2018)

9. Gusev, M.: Dew computing architecture for cyber-physical systems and IoT. Internet Things **11**, 100186 (2020)

10. Gusev, M.: Serverless and deviceless dew computing: founding an infrastructureless computing. In: 2021 IEEE 45th Annual Computers, Software, and Applications Conference (COMPSAC), pp. 1814–1818. IEEE (2021)

11. Gusev, M.: What makes dew computing more than edge computing for internet of things. In: 2021 IEEE 45th Annual Computers, Software, and Applications Conference (COMPSAC), pp. 1795–1800. IEEE (2021)

12. Gusev, M.: AI cardiologist at the edge: a use case of a dew computing heart monitoring solution. In: Artificial Intelligence and Machine Learning for EDGE Computing, pp. 469–477. Elsevier (2022)

13. Gusev, M.: Scalable dew computing. Appl. Sci. **12**(19), 9510 (2022)

14. Gusev, M., Dustdar, S.: Going back to the roots: the evolution of edge computing, an IoT perspective. IEEE Internet Comput. **22**(2), 5–15 (2018)

15. Kim, T., Yoo, S.E., Kim, Y.: Edge/fog computing technologies for IoT infrastructure. Sensors **21**(9), 3001 (2021)

16. Klonoff, D.C.: Fog computing and edge computing architectures for processing data from diabetes devices connected to the medical internet of things. J. Diabetes Sci. Technol. **11**(4), 647–652 (2017)

17. Lynn, T., Mooney, J.G., Lee, B., Endo, P.T.: The cloud-to-thing continuum: opportunities and challenges in cloud, fog and edge computing (2020)

18. Mao, Y., You, C., Zhang, J., Huang, K., Letaief, K.B.: A survey on mobile edge computing: the communication perspective. IEEE Commun. Surv. Tutor. **19**(4), 2322–2358 (2017)

19. Medhi, K., Ahmed, N., Hussain, M.I.: Dew-based offline computing architecture for healthcare IoT. ICT Express **8**(3), 371–378 (2022)

20. Nastic, S., et al.: A serverless real-time data analytics platform for edge computing. IEEE Internet Comput. **21**(4), 64–71 (2017)

21. Pan, Y., Thulasiraman, P., Wang, Y.: Overview of cloudlet, fog computing, edge computing, and dew computing. In: Proceedings of The 3rd International Workshop on Dew Computing, pp. 20–23 (2018)

22. Ray, P.P.: An introduction to dew computing: definition, concept and implications. IEEE Access **6**, 723–737 (2017)

23. Ray, P.P., Dash, D., De, D.: Edge computing for internet of things: a survey, e-healthcare case study and future direction. J. Netw. Comput. Appl. **140**, 1–22 (2019)

24. Remy, S.L., Gajananan, K., Karve, A.: Redefining edge computing. In: 2022 IEEE Cloud Summit, pp. 113–117. IEEE (2022)

25. Rindos, A., Wang, Y.: Dew computing: the complementary piece of cloud computing. In: 2016 IEEE International Conferences on Big Data and Cloud Computing (BDCloud), Social Computing and Networking (SocialCom), Sustainable Computing and Communications (SustainCom)(BDCloud-SocialCom-SustainCom), pp. 15–20. IEEE (2016)

26. Roy, A., Midya, S., Majumder, K., Phadikar, S.: Distributed resource management in dew based edge to cloud computing ecosystem: a hybrid adaptive evolutionary approach. Trans. Emerg. Telecommun. Technol. **31**(8), e4018 (2020)

27. Satyanarayanan, M.: Edge computing: vision and challenges. USENIX Association, Santa Clara, USA (2017)

28. Satyanarayanan, M., Bahl, P., Caceres, R., Davies, N.: The case for VM-based cloudlets in mobile computing. IEEE Pervasive Comput. **8**(4), 14–23 (2009)

29. Shi, W., Cao, J., Zhang, Q., Li, Y., Xu, L.: Edge computing: vision and challenges. IEEE Internet Things J. **3**(5), 637–646 (2016)

30. Singh, S.P., Nayyar, A., Kumar, R., Sharma, A.: Fog computing: from architecture to edge computing and big data processing. J. Supercomput. **75**(4), 2070–2105 (2019)

31. Skala, K., Davidovic, D., Afgan, E., Sovic, I., Sojat, Z.: Scalable distributed computing hierarchy: cloud, fog and dew computing. Open J. Cloud Comput. (OJCC) **2**(1), 16–24 (2015)

32. Šojat, Z., Skala, K.: The dawn of dew: dew computing for advanced living environment. In: 2017 40th International Convention on Information and Communication Technology, Electronics and Microelectronics (MIPRO), pp. 347–352. IEEE (2017)

33. Šojat, Z., Skala, K.: The rainbow: integrating computing into the global ecosystem. In: 2019 42nd International Convention on Information and Communication Technology, Electronics and Microelectronics (MIPRO), pp. 222–229. IEEE (2019)

34. Stojmenovic, I., Wen, S.: The fog computing paradigm: scenarios and security issues. In: 2014 Federated Conference on Computer Science and Information Systems (FedCSIS), pp. 1–8. IEEE (2014)

35. Wang, Y.: Cloud-dew architecture. Int. J. Cloud Comput. **4**(3), 199–210 (2015)
36. Wang, Y.: Definition and categorization of dew computing. Open J. Cloud Comput. (OJCC) **3**(1), 1–7 (2016)
37. Wang, Y., Thulasiraman, P.: Post-cloud computing models and their comparisons. In: Zhang, Q., Wang, Y., Zhang, L.-J. (eds.) CLOUD 2020. LNCS, vol. 12403, pp. 141–151. Springer, Cham (2020). https://doi.org/10.1007/978-3-030-59635-4_10
38. Yahuza, M., et al.: Systematic review on security and privacy requirements in edge computing: State of the art and future research opportunities. IEEE Access **8**, 76541–76567 (2020)
39. Zhou, Y., Zhang, D., Xiong, N.: Post-cloud computing paradigms: a survey and comparison. Tsinghua Sci. Technol. **22**(6), 714–732 (2017)

# E-learning and E-services

# Unveiling Insights: Analyzing Application Logs to Enhance Autism Therapy Outcomes

Bojan Ilijoski[✉] and Nevena Ackovska

Faculty of Computer Science and Engineering, Ss Cyril and Methodiuos University,
Skopje, North Macedonia
{bojan.ilijoski,nevena.ackovska}@finki.ukim.mk

**Abstract.** Leveraging advancements in information technology and the inherent interest of children with autism in robots and technology, this study explores the crucial role of analyzing application logs in enhancing therapy experiences for children with autism. By examining these logs, valuable insights can be obtained, enabling performance tracking, evidence-based evaluation, personalization of interventions, and continuous improvement. This will allow us to get more information about children's preferences and behavior even when we are not in direct contact with them, by extending onsite robot therapies to the home environment. This research contributes to the understanding of the transformative power of log analysis and its implications for optimizing therapy experiences and advancing treatment for children with autism.

**Keywords:** autism spectrum disorder · human-robot interaction · human-computer interaction · logs · usability

## 1 Introduction

In the last decade, the field of robotics has made substantial strides, witnessing progress across various domains. Initially, industries and service sectors were at the forefront of robot adoption. However, there has been a recent surge in personal and caregiving robots, with their numbers now approaching those of service robots. This upsurge in personal robots has created a growing demand for trained individuals who can oversee interactions between humans and robots (HRI scenarios). Social robots, in particular, have proven highly effective in specific HRI contexts. They excel in assisting individuals with disabilities and providing socially assistive therapy for children with autism [7]. Autism Spectrum Disorder (ASD) is a complex neurodevelopmental condition characterized by difficulties in social communication, interaction, and repetitive behaviors. With the increasing prevalence of ASD diagnoses, there is an urgent need to develop

---

Supported by Faculty of Computer Science and Engineering, Skopje, N. Macedonia.

M. Mihova and M. Jovanov (Eds.): ICT Innovations 2023, CCIS 1991, pp. 111–124, 2024.
https://doi.org/10.1007/978-3-031-54321-0_8

intervention methods and treatments tailored to the unique requirements of children with ASD. Early interventions have demonstrated significant benefits and improved outcomes for these children, which has spurred continuous efforts to innovate new therapeutic approaches [16]. Robot-assisted therapies for children with autism have shown promise, with participants displaying keen interest and active engagement. The integration of robots into therapy sessions stimulates interaction and curiosity, thereby facilitating learning and skill development in a safe and supportive environment. Notably, children with autism often exhibit a strong affinity for technological devices, including robots, making them valuable tools in their treatment. By harnessing these technological advancements, the incorporation of robots and applications can augment conventional therapy methods for children with autism [21].

In this paper, we present a case study that explores the benefits of collecting and analyzing logs from a mobile and web application that extends onsite robot therapies for children with autism into the home environment. The intervention took place at a Children's Hospital, where the sessions involved the Kaspar robot [5], positioned on a table in front of the children and a therapist controlled the robot remotely using a controller or computer. The therapy sessions included various scenarios such as imitation games, learning emotions, turn-taking games, learning animals and animal sounds, learning sounds and words, hygiene, and food. Kaspar effectively engaged children with autism spectrum disorder (ASD) through a range of therapeutic and educational games, including turn-taking, joint attention, collaborative games, cause and effect games, and more. The robot was able to hold objects, respond to touches, and perform specific movements, enhancing the interactive experience for the children.

The application serves as a means for ASD children to replicate their therapy activities outside of pediatric institutions. It represents a kind of mirroring of the whole pre-assessment through videos of the scenarios with Kaspar that the children practice during the therapies [9]. By examining the advantages of analyzing application logs and harnessing the potential of technology, we aim to contribute to the broader understanding of how digital applications can positively impact therapy experiences for children with autism and open new possibilities for their treatment and development. One major benefit of analyzing application logs is the ability to track and monitor the performance of application that extends therapy sessions or interventions for children with autism. Detailed logs provide professionals with insights to assess and fine-tune therapy strategies, adapt interventions to individual needs, and identify progress patterns. The analysis of logs allows for quantitative evaluation, objective measurement of outcomes, and comparison of therapy effectiveness. It also enables personalized and adaptive interventions, optimizing engagement and learning outcomes. Analyzing logs facilitates continuous improvement in application design and functionality, addressing usability issues and enhancing user experience.

This paper is organized as follows. In the second chapter we present the current achievements in HCI for people with autism. In the third chapter we introduce the method of collecting application logs. The fourth chapter presents the analysis of the collected and discuses the obtained results, and the last chapter elaborates on the conclusion of the research.

## 2 Related Work

The use of applications in interventions for children with autism could be observed in various areas such as communication, collaboration, language, and social skills. These applications, often in the form of serious games, are designed for personal computers, tablets, and mobile phones, aiming to enhance education and therapy [10,17]. They typically offer user-friendly interfaces and can use artificial intelligence capabilities for improved outcomes [12,15,18]. Some applications focus on emotion recognition, while others utilize phone sensors for data extraction and analysis [11,13,23]. Additionally, the inclusion of robots in autism therapy has shown promise, especially in social integration and development. Various types of robots, such as humanoid and toy/animal-shaped ones, are used in human-robot interaction therapy and these robots provide a secure and engaging environment for children to learn and practice new skills [3,4,14,20]. While robot-based therapies have shown potential, the challenge lies in practicing the acquired skills beyond clinical settings. Therefore, the integration of applications as a complementary tool outside the clinic is very important. Several research studies have investigated the use of application logs and technology in augmenting therapy for children with autism, particularly in simulating onsite robot therapy activities. Tanaka et al. [22] conducted a study involving the use of a virtual reality system with haptic feedback, enabling children with autism to engage in social interactions with a virtual character. The results demonstrated improved social communication skills and increased engagement. Additionally, Diehl et al. [6] investigated the effectiveness of a tablet-based application designed to enhance joint attention skills in children with autism. The study demonstrated significant improvements in joint attention and social interaction abilities. Moreover, research by Anzalone et al. [2] focused on the use of mobile applications for sensory integration therapy in children with autism. The applications provided interactive sensory experiences and visual aids to facilitate therapy sessions. The results indicated improved sensory processing and functional skills in the participants. These studies illustrate the potential benefits of using application logs and technology to replicate and enhance therapy experiences for children with autism.

The work presented in this paper builds on our previous research related to mirroring and extending robot therapy for children with autism through a mobile application, as well as our long-term work with the development of applications intended for children with autism. We analyzed the entire therapy process in clinical settings and based on that we built a web and mobile application. This application allowed the robot therapy that children practiced in clinical settings to be continued at home. The results obtained from these studies are promising. [1,8,9,24].

# 3    Methods

Logs are a common practice for tracking user behavior when using applications. This data allows researchers to observe user activity in near real time and capture actual user behavior rather than recalled behaviors or subjective impressions. Unlike laboratory studies that take place in a controlled environment and may not be truly representative of the user's behavior outside, logs represent the user's natural behavior without the influence of testers and observers. Additionally, the group observed in this way can be much larger and more diverse. This process includes three stages of collecting, cleaning and analyzing user logs. The collection of logs is usually done through modules installed in the application itself or external tools. The data included in the logs can be of different types, but most often it boils down to the time when the event occurred, the user who created it, and information about the event itself. Depending on the platform, application type, and experiments, additional parameters and conditions may be included. In order to take advantage of the logs, they were also implemented in our application. In our case, a kind of case study and experimental methodology was used, where we try to do an in-depth exploration of a few cases to gain insights through controlled experiments with the aim of analyzing multiple cases for comparison and contrast.

As already mentioned in the previous paragraphs, an application was built to support robot therapy. The purpose of this application was to overcome the resource limitations that existed in clinical therapy i.e. only one robot and a small number of trained staff to work with it. Because of these limitations, a small number of children had access to a limited number of therapies, which is one of the main reasons to build an application that supports this process. The app actually contains videos of the scenarios that the children practice with the robot Kaspar during the sessions and they can be played in a simple way by clicking on a button with an icon from the scenario.

```
{
  "dateTime": "2020-09-04T08:25:46.017Z",
  "data": {
    "click": {
      "x": 92.3076923076923,
      "y": 22.129783693843592
    },
    "message": {
      "key": "playVideo",
      "value": "mk/Hi, I'm Kaspar.mp4"
    }
  }
}
```

**Fig. 1.** Log example

In our specific case, we integrated a logger into the application to closely monitor users' interactions. In this endeavor, we successfully collected activities from six different devices. These logs are stored in .json format and exhibit varying sizes, ranging from 6.9 KB to 1.4 MB. Importantly, these logs are obtained directly from the device with explicit parental consent. Each .json file corresponds to a specific device, which is essentially one child's usage data. Within these logs, each record signifies an application event. An event can be a user's click (touch) or the launching of an application. For each event, we capture the timestamp (dateTime) when it occurred, the position on the screen where it transpired (data.click), and the type of action performed (data.massage). You can find an example of such a log in Fig. 1.

To streamline the implementation of our analyses, we have restructured the data from the JSON format into a tabular format. Additionally, we've introduced derived attributes, such as session names and event durations. Event duration represents the time elapsed between two consecutive events. It's worth noting that the number of logs per file can vary significantly. This variance is expected, considering that the application is utilized with varying frequencies by different users. For the sake of more meaningful analysis, we've organized these events into sessions. A session denotes a continuous period of using the application. It's important to clarify that we don't consider the time between two application launches as a session since the application might remain active in the background without any usage. We've observed that if we measure the duration of a session from the beginning of an application's use to the last event before a new session begins, some sessions can span up to 14 days. This prolonged duration can occur when the app remains running in the background. To ensure the relevance of our sessions, we define a session as concluded if there is a 5-minute period without any activity, meaning no events are registered during this interval.

In order to be able to collect the data and analyze it, we asked for and received permission from all the parents of the children who were included in the study. What's more, there is no data transfer over the network in the whole process. The application works completely offline and in order to take the data of the logs, it is necessary for the parents to give us the device on which the application is installed and thus we can extract the data. Before the analysis, the data were completely anonymized.

## 4   Results

There is a total of 17,840 events recorded, including 370 application launches and 538 distinct sessions. Within the context of session initiation, it is valuable to examine both the quantity of events and the duration of these sessions. Typically, sessions encompass a range of events, with most falling between 0 and 100 events. On average, a session comprises 33 events, although it's important to note that outliers exist. To provide a more detailed picture, it's worth highlighting that 75% of sessions have fewer than 53 events. Another noteworthy observation is that a quarter of the sessions consist of three or fewer events. This indicates instances where the application was launched but experienced minimal usage. To be precise, there are 107 sessions with just one or two events. Turning our attention to session durations, roughly half of them last between 0 and 5 min. However, there are sessions that extend up to 25 min. This diversity in session durations underscores the variability in user engagement with the application.

Figure 2 presents a scatter plot of the duration and number of events per session. The x-axis represents the duration of sessions in minutes, while the y-axis represents the number of events for that session (the definitions of sessions and events are given above). As we can see from the figure no correlation can be observed between them. Some outliers can also be noticed by these two attributes, i.e. that we have a session with many events (more than 175) and a small duration (less than a minute), as well as sessions with a large duration ($\approx 20$ min) but a small number of events ($<25$).

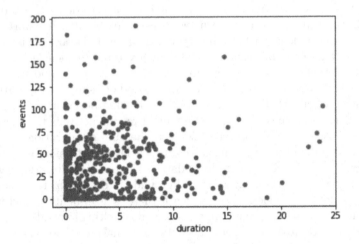

**Fig. 2.** Scatter plot of number of event and events duration

Figure 3 shows the clicks on the application. As expected, most of them are in the right part, in the part with buttons where the scenario selection is actually made and the only active part i.e. part where a click causes an action. Clicks can also be observed in the inactive part and two interesting clusters of clicks that correspond to the position of the thumbs holding the tablet. There is also a certain number of clicks on the edges which can perhaps also be attributed to a similar pressure when holding or moving the tablet itself. Another thing that can be noticed is the accuracy of the users i.e. most of the active clicks are located in the middle of the buttons.

Figure 4 shows the number of sessions per hour. What can be seen here are three intervals that bounce around the frequency of use. These are from 8 am to 11 am, from 2 pm to 3 pm and from 7 pm to 8 pm. In the period from 10 pm to 5 am there are almost no sessions, which is to be expected since the application is used by children.

Analysis of application usage logs can produce significant information about users, their behaviors, application usage patterns, etc. In the following, we will consider several aspects related to log analysis. The first is the app's click-through rate. An increased click rate can mean that the application is not

**Fig. 3.** Clicks on the application

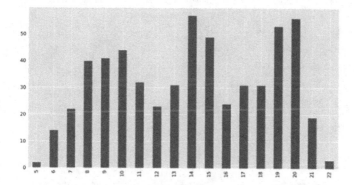

**Fig. 4.** Bar chart of number of sessions grouped by hour of day

responding fast enough, so the user repeats the action again and again, that the user is not interested in the scenarios and does not use the application, but just clicks on the screen in random places, or that (s)he is only interested in the same scenario and plays it over and over again, etc.

Figure 5 shows the scenarios and the timestamp when they were run for four different sessions. Four different patterns have been deliberately chosen. The first one presents a session where the user is playing all scenarios one by one. Furthermore the scenarios are played in the way as they are shown in application under certain category, then the category is changed and the users play the scenarios from the selected category and so on. This looks like exploring the application and all scenarios presented there. The next one presents playing scenarios, some of them multiple times, then switch the category and continue with playing scenarios from that category. The third one presents playing scenarios multiple times from one category and the last one play one scenario over and over again. Specifically for the session in the fourth image, we additionally check that half of the clicks were in a random position with no action, and half were

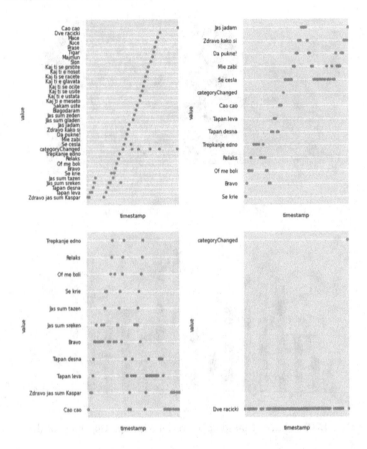

**Fig. 5.** Four random sessions. Different scenarios on y axes and timestamp on x axes.

launching the "Two Hands" scenario. The time difference between two launches was to short which means that the user didn't wait for the scenario to finish.

In Fig. 6 we have the summary view of the number of consecutive empty clicks for all sessions. As an empty click is considered a click that does not cause an action, i.e. click that is outside of a button for scenario playing. It can be noticed that for some of the users we have more than 10 consecutive empty clicks and we even have a case with more than 20 times clicked on empty.

The Generalized Sequence Pattern (GSP) algorithm [19] was applied in order to detect repeating patterns of the scenarios. With a confidence parameter value of 0.3 over the entire data set, the following frequency sequences were obtained (the best 5 from each group) (Fig. 7).

In order to get a better picture of the processes i.e. for the sequence of scenarios when using the application, we can represent them as a directed graph, where the vertices are scenarios, and there is a directed edge between them if they often occur one after the other (Fig. 8). The thickness of the ribs is directly proportional to the frequency. For the creation of this graph, a threshold was

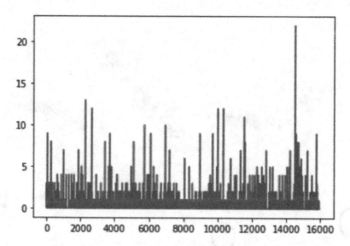

**Fig. 6.** Number of empty clicks per session (The x-axis represents the sessions, for all patients sorted by start time, the y-axis the number of empty clicks).

```
Length 1:
('Category changed', ): 194
('Hi, I'm Kaspar',): 173
('Blink once',): 172,
('Bye Bye',): 170
('Drum left',): 156,
Length 2:
('Category changed', 'Two hands'): 136,
('Bravo', 'Oh, it hurts me'): 130
('Drum left', 'Drum right'): 128
('I'm happy', 'I'm sad'): 121
('Combs hair', 'Brushes teeth'): 120
Length 3:
('Combs hair', 'Brushes teeth', 'High five!'): 99
('I'm happy', 'I'm sad', 'Hiding'): 96
('Category changed', 'Combs hair', 'Brushes teeth'): 95
('Drum left', 'Drum right', 'I'm happy'): 93
('Hiding', 'Bravo', 'Ohh, it hurts me'): 91
```

**Fig. 7.** Output of GSP analysis

used for the edges, that is, the edges with over 150 consecutive selections are shown. This is chosen experimentally and is intended to show only the frequency relationships. The size of the threshold depends on how the graph will look. A larger threshold leads to a sparse graph, and a smaller one leads to a graph in which there is a connection between almost all vertices. In Fig. 8, it can be observed that six loosely connected components are formed that correspond to the categories. This behavior is also to be expected because typically users select a category and then stick to playing the scripts that are in that category.

In Fig. 9 shown with color map are the number of consecutive scenarios. The scenarios are arranged in the order they appear in the app, so what can be observed as a trend is that users generally want to repeat the same scenario or play the next or previous one. Clusters of several scenarios can be observed, and these are generally the same as those detected previously.

These analyzes can help us with some kind of notifications when anomalies and not typical behaviors are observed in the system, and can also serve us

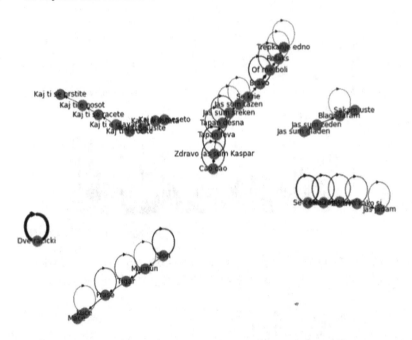

**Fig. 8.** Directed graph from played scenarios.

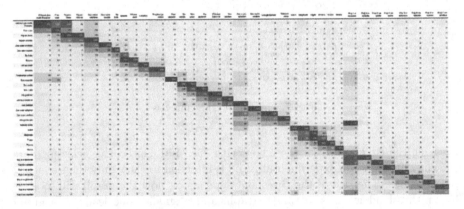

**Fig. 9.** Matrix of number of sequentially playing of scenarios.

to cluster similar behaviors between patients, such as finding similar sequences of events (scenario sessions). For this purpose, clustering was done with several different techniques, but the most impressive results were observed with the DBSCAN algorithm, where a metric derived from Levenshtein distance was used as a measure of distance between sequences. So the distance between two sequences a and b is calculated as

$$
lev_{a,b}(i,j) = \begin{cases} max(i,j) & \text{if } min(i,j) = 0 \\ min \begin{cases} lev_{a,b}(i-1,j)+1 \\ lev_{a,b}(i,j-1)+1 \\ lev_{a,b}(i-1,j-1)+1_{(a_i \neq b_j)} \end{cases} & \text{otherwise} \end{cases} \tag{1}
$$

where $a_i$ and $b_j$ are the $i - th$ event of the sequence (session) $a$ and the $j - th$ event of the sequence $b$. This distance is normalized to rank $[0, 1]$ by dividing by $max(len(a), len(b))$. Experiments were made with different values for $eps$ and $min\_samples$ parameters in the range of $(0, 1]$ and $[2, 7]$. Expected for high values of the $eps$ parameter, the number of clusters and outliers is small, because all sequences fall into one cluster. On the other hand, small values for this parameter contribute to a large number of outliers. The $min\_samples$ parameter has a greater influence on the number of clusters, but the $eps$ parameter also plays a significant role. Here it can be noted that for small values of $min\_samples$ we generally get a large number of clusters, of course depending on the value of the $eps$ parameter.

Depending on the values of the parameters, the number of clusters and types of sequences that occur are different, but in general some patterns can be distinguished. After analyzing the number of clusters, the number of outliers and the number of sequences, twenty clustering were considered. What can be observed from them is that the number of outliers generally ranges between 230 and 350, and the number of clusters between 4 and 15. In those clustering where we have values for $eps \leq 0.5$, two larger groups of clusters with 70 each are observed, up to 100 sequences and several smaller groups of 2 to 20 sequences. For values of $eps < 0.5$, one larger group and several smaller ones are distinguished. If we observe the clusters according to users, in some of the clustering we can notice that we have sequences belonging to different users, which can be an indicator of a similar way of using the application between those users.

## 5   Discussion and Conclusion

The benefits of the application itself are great, especially because it allows the children to somehow continue the therapy in conditions when they cannot be in a clinic. The analysis of the logs resulting from it and the information obtained from them can be multifaceted. Regarding the development of the application itself, this information can contribute to obtaining conclusions about how it is made, how users use it, what errors occur, whether the actions taken are expected, whether the user is comfortable with the user interface, etc. This can lead to improving the system, adapting it to users, faster learning, encouraging and improving the overall user experience. It can also be beneficial for parents who can track their child's behavior and interests, and finally for therapists who can track a child's behavior, looking for similarities between behaviors, differences between sessions, progress, and so on.

But this approach also has drawbacks. The information that is collected relates to what users do, but not why they do it and whether they are satisfied.

Furthermore, it is extremely difficult to detect the real user, because the device may be shared between multiple users using the application, or in our case, we cannot know if the child is using the application independently, with the help of a parent, or the parent is the one who uses it. Also a real challenge is the definition of metrics for evaluation of the experiments, due to the different way of using the application for different users. Also different metrics can lead to different results for different users. If all of this is taken into account when analyzing logs, it can produce significant information about the application and users. Also, the group of children that was observed is small, so it is difficult to draw any general conclusions based on the data analysis.

In general, log collection in applications for children with autism can be very useful. Many of these children are non-verbal so this is probably the only way to see and gauge their interest. Also, this analysis contributed to finding some patterns of application usage within one session with the same or different users. These patterns can further provide us with information about how the application is used and capture some repetitive user actions that are specific to users with autism. This also allow us to find out which scenarios they are interested in, which can help us to implement a higher degree of personalization in the later stages. In this way, we can track the child's progress over time and how his needs and interests change for the scenarios in the application. Overall log analysis from applications for children with autism provides valuable insights into their performance, enables evidence-based evaluation of therapy outcomes, facilitates personalized interventions, and drives continuous improvement and innovation in application design and functionality. This knowledge empowers professionals to optimize therapy approaches and enhance the overall support and outcomes for children with autism.

**Acknowledgements.** This work was partially financed by the Faculty of Computer Science and Engineering at the Ss. Cyril and Methodius University in Skopje.

# References

1. Ackovska, N., Ilijoski, B.: Online availability of a robot therapy for children with autism. In: SoutheastCon 2023, pp. 603–609 (2023). https://doi.org/10.1109/SoutheastCon51012.2023.10115087
2. Anzalone, S.M., et al.: How children with autism spectrum disorder behave and explore the 4-dimensional (spatial 3D+time) environment during a joint attention induction task with a robot. Res. Autism Spectrum Disord. 8(7), 814–826 (2014). https://doi.org/10.1016/j.rasd.2014.03.002, https://www.sciencedirect.com/science/article/pii/S1750946714000452
3. Bartl-Pokorny, K., et al.: Robot-based intervention for children with autism spectrum disorder: a systematic literature review. IEEE Access 1 (2021). https://doi.org/10.1109/ACCESS.2021.3132785
4. Costa, A.P., et al.: More attention and less repetitive and stereotyped behaviors using a robot with children with autism. In: 2018 27th IEEE International Symposium on Robot and Human Interactive Communication (RO-MAN), pp. 534–539 (2018). https://doi.org/10.1109/ROMAN.2018.8525747

5. Dautenhahn, K., et al.: KASPAR - a minimally expressive humanoid robot for human-robot interaction research. Appl. Bionics Biomech. **6**, 369–397 (2009). https://doi.org/10.1155/2009/708594

6. Diehl, J.J., Schmitt, L.M., Villano, M., Crowell, C.R.: The clinical use of robots for individuals with autism spectrum disorders: a critical review. Res. Autism Spectrum Disord. **6**(1), 249–262 (2012). https://doi.org/10.1016/j.rasd.2011.05.006, https://www.sciencedirect.com/science/article/pii/S1750946711000894

7. Fachantidis, N., Syriopoulou-Delli, C.K., Zygopoulou, M.: The effectiveness of socially assistive robotics in children with autism spectrum disorder. Int. J. Dev. Disabil. **66**(2), 113–121 (2018)

8. Ilijoski, B., Ackovska, N.: Developing applications for children with special needs into a project based learning approach at human - computer interaction course. In: 2022 IEEE Global Engineering Education Conference (EDUCON), pp. 1322–1329 (2022). https://doi.org/10.1109/EDUCON52537.2022.9766483

9. Ilijoski, B., Ackovska, N., Zorcec, T., Popeska, Z.: Extending robot therapy for children with autism using mobile and web application. Sensors **22**(16) (2022). https://doi.org/10.3390/s22165965, https://www.mdpi.com/1424-8220/22/16/5965

10. Iyer, S., Mishra, R.S., Kulkami, S.P., Kalbande, D.: Assess autism level while playing games. In: 2017 2nd International Conference on Communication Systems, Computing and IT Applications (CSCITA), pp. 42–47 (2017). https://doi.org/10.1109/CSCITA.2017.8066573

11. Kalantarian, H., Jedoui, K., Washington, P., Wall, D.P.: A mobile game for automatic emotion-labeling of images. IEEE Trans. Games **12**(2), 213–218 (2020). https://doi.org/10.1109/TG.2018.2877325

12. Kołakowska, A., Landowska, A., Anzulewicz, A., Sobota, K.: Automatic recognition of therapy progress among children with autism. Sci. Rep. **7** (2017). https://doi.org/10.1038/s41598-017-14209-y

13. Kołakowska, A., Landowska, A., Karpienko, K.: Gyroscope-based game revealing progress of children with autism. In: Proceedings of the 2017 International Conference on Machine Learning and Soft Computing, ICMLSC 2017, pp. 19–24. Association for Computing Machinery, New York (2017). https://doi.org/10.1145/3036290.3036324

14. Kouroupa, A., Laws, K.R., Irvine, K., Mengoni, S.E., Baird, A., Sharma, S.: The use of social robots with children and young people on the autism spectrum: a systematic review and meta-analysis. PLoS One **17**(6), e0269800 (2022)

15. Nazardeen, N., Yapa, L., Imthath, M., Pathirana, S., Jayathunga, P.: A software package for analyzing and evaluating children with autism spectrum disorder (ASD) in Sri Lanka, pp. 985–990 (2020). https://doi.org/10.1109/DASA51403.2020.9317181

16. Park, H.R., et al.: A short review on the current understanding of autism spectrum disorders. Exp. Neurobiol. **25**(1), 1–13 (2016)

17. Pistoljevic, N., Hulusic, V.: An interactive e-book with an educational game for children with developmental disorders: a pilot user study. In: 2017 9th International Conference on Virtual Worlds and Games for Serious Applications (VS-Games), pp. 87–93 (2017). https://doi.org/10.1109/VS-GAMES.2017.8056575

18. Popescu, A.L., Popescu, N.: Machine learning based solution for predicting the affective state of children with autism. In: 2020 International Conference on e-Health and Bioengineering (EHB), pp. 1–4 (2020). https://doi.org/10.1109/EHB50910.2020.9280194

19. Prado Lima, J.A.d.: GSP-Py - generalized sequence pattern algorithm in Python (2020). https://doi.org/10.5281/zenodo.3333987

20. Robins, B., Dautenhahn, K., Dickerson, P.: From isolation to communication: a case study evaluation of robot assisted play for children with autism with a minimally expressive humanoid robot, vol. Cancun, pp. 205–211 (2009). https://doi.org/10.1109/ACHI.2009.32

21. Szymona, B., et al.: Robot-assisted autism therapy (RAAT). Criteria and types of experiments using anthropomorphic and zoomorphic robots. Review of the research. Sensors (Basel) **21**(11), 3720 (2021)

22. Tanaka, J.W., et al.: Using computerized games to teach face recognition skills to children with autism spectrum disorder: the let's face it! Program. J. Child Psychol. Psychiatry **51**(8), 944–952 (2010)

23. Washington, P., et al.: Improved digital therapy for developmental pediatrics using domain-specific artificial intelligence: machine learning study (preprint). JMIR Pediatr. Parent. **5** (2020). https://doi.org/10.2196/26760

24. Zorcec, T., et al.: Enriching human-robot interaction with mobile app in interventions of children with autism spectrum disorder. PRILOZI **42**(2), 51–59 (2021). https://doi.org/10.2478/prilozi-2021-0021

# Image Processing

# Semantic Segmentation of Remote Sensing Images: Definition, Methods, Datasets and Applications

Vlatko Spasev[(⊠)], Ivica Dimitrovski, Ivan Kitanovski, and Ivan Chorbev

Faculty of Computer Science and Engineering, Ss Cyril and Methodiuos University,
Skopje, North Macedonia
{vlatko.spasev,ivica.dimitrovski,ivan.kitanovski,
ivan.chorbev}@finki.ukim.mk

**Abstract.** Semantic segmentation of remote sensing images is a vital task in the field of remote sensing and computer vision. The goal is to produce a dense pixel-wise segmentation map of an image, where a specific class is assigned to each pixel, enabling detailed analysis and understanding of the Earth's surface. This paper provides an overview of semantic segmentation in remote sensing, starting with a definition of the task and its significance in extracting valuable information from remote sensing imagery. Various methods used for semantic segmentation in remote sensing are discussed, including traditional approaches such as region-based and pixel-based methods, as well as more recent deep learning-based techniques. Next, the paper delves into the available datasets for semantic segmentation of remote sensing images. Many available datasets are reviewed, highlighting their characteristics, including the number of images, image size, number of labels, spatial resolution, format and spectral bands. These datasets serve as valuable resources for training, evaluating, and benchmarking semantic segmentation algorithms in remote sensing applications. Furthermore, the paper highlights the broad range of applications enabled by semantic segmentation in remote sensing, including urban planning, land cover mapping, disaster management, environmental monitoring, and precision agriculture. Overall, this paper serves as a comprehensive guide to semantic segmentation of remote sensing images, providing insights into its definition, methods, available datasets and wide-ranging applications.

**Keywords:** Remote Sensing Images · Semantic Segmentation · Deep Learning · Earth Observation

## 1 Introduction

Remote sensing images refer to images captured from a distance by sensors or instruments mounted on satellites, aircraft, drones, or other platforms. These images are used to collect information about the Earth's surface, atmosphere,

Supported by Faculty of Computer Science and Engineering, Skopje, N. Macedonia.

M. Mihova and M. Jovanov (Eds.): ICT Innovations 2023, CCIS 1991, pp. 127–140, 2024.
https://doi.org/10.1007/978-3-031-54321-0_9

and other objects or phenomena without direct physical contact [44]. In recent years, the advent of sophisticated machine learning techniques, coupled with the abundance of remote sensing data, has paved the way for significant advancements in image analysis and interpretation [10, 18].

The concept of semantic segmentation has made substantial strides [21]. Its application to remote sensing imagery spans various domains and has been a prominent research area for decades [47]. Operating at the forefront of computer vision, semantic segmentation equips machines with the capability to intricately understand and demarcate image content down to individual pixels. Unlike traditional object detection methods that label entire objects or regions within an image, semantic segmentation meticulously labels each pixel according to its associated object or class. This finer granularity of analysis endows us with a deeper understanding of the intricate spatial distribution of features within remote sensing images.

The implications of semantic segmentation within the realm of remote sensing are vast and profound. It finds application in a multitude of fields, including urban planning, agriculture, environmental monitoring, disaster management, forestry, and more [31]. Semantic segmentation enables the automated extraction of vital information from imagery, unraveling patterns and changes that might otherwise elude human perception. By unraveling the complex tapestry of pixels, semantic segmentation unveils insights that drive informed decision-making and facilitate holistic comprehension of the Earth's ever-evolving landscapes.

This paper makes significant contributions to the understanding and advancement of semantic segmentation within the context of remote sensing imagery. It provides a comprehensive and cohesive overview of the concept of semantic segmentation in the domain of remote sensing. It serves as an accessible introduction for both novice and seasoned researchers, offering a clear understanding of the underlying principles and significance of semantic segmentation in interpreting remote sensing data. A significant contribution of the paper lies in its exploration of various methods and techniques employed in semantic segmentation for remote sensing images. The paper conducts a thorough examination of datasets used in training and evaluating semantic segmentation models for remote sensing imagery. Highlighting various applications, the paper demonstrates the real-world implications of semantic segmentation within the realm of remote sensing.

In the reminder, we first explain the main characteristics of the remote sensing image data and then define the task of semantic segmentation, outlining the input data and the desired output. Various machine learning methods used for semantic segmentation are discussed, along with evaluation measures to assess the performance of trained models. The paper further summarizes and highlights different datasets available for training and evaluating semantic segmentation models for remote sensing data. Lastly, the paper explores the potential applications of semantic segmentation in the context of remote sensing data. It discusses how semantic segmentation can be employed in various domains such as urban planning, disaster management, environmental monitoring, precision agriculture, deforestation analysis, climate assessment, and water resource management. These applications showcase the broad range of benefits that semantic segmentation offers in understanding and analyzing remote sensing imagery.

## 2    Characteristics of Remote Sensing Image Data

Remote sensing images can be broadly categorized into two main types: aerial images and satellite images. Satellite images and aerial images are both valuable sources of remote sensing data, but they differ in how they are acquired and the characteristics of the imagery. Remote sensing data encompasses various aspects of information representation, including spectral, spatial, radiometric, and temporal resolutions. Spectral resolution involves the bandwidth and sampling rate employed for data capture. High spectral resolution signifies narrower bands of the spectrum, while low resolution indicates broader bands. These spectral bands span diverse wavelengths such as ultraviolet, visible, near-infrared, infrared, and microwave. Image sensors range from multi-spectral, covering numerous bands (e.g., Sentinel-2[1] with 12 bands), to hyper-spectral sensors like Hyperion (part of the EO-1 satellite), gathering thousands of spectral bands (0.4–2.5 μm) [34].

Spatial resolution refers to the Earth's surface area represented by each pixel in an image. Higher spatial resolutions (small pixel size) capture finer details, whereas lower resolutions (large pixel size) retain fewer details. Moderate Resolution Imaging Spectroradiometer (MODIS), for instance, observes most bands with a spatial resolution of 1 km, where each pixel signifies a 1 km × 1 km ground area [19]. Conversely, UAV-captured images can achieve highest spatial resolutions, even less than 1 cm pixel size [33].

Radiometric resolution defines the sensor's capability to record signals of varying strengths (dynamic range). A larger dynamic range enables the detection of intricate details in recordings. Landsat 7 records 8-bit images, discerning 256 distinct gray values of reflected energy, while Sentinel-2 boasts a 12-bit radiometric resolution (4095 gray values). Enhanced radiometric resolution facilitates the differentiation of subtle variations in ocean color, crucial for water quality assessment.

Temporal resolution denotes how often a satellite revisits a specific observation area. Polar-orbiting satellites exhibit varying temporal resolutions, ranging from 1 to 16 days (e.g., ten days for Sentinel-2). Temporal considerations are pivotal in monitoring changes within observation areas, encompassing aspects like land use alteration, deforestation, and mowing.

Satellite images are captured by sensors mounted on satellites orbiting the Earth. These satellites can be classified into different types, including optical, radar, and thermal satellites [44]. Satellites, with varying altitudes and predetermined orbits, capture images across large expanses at regular intervals. Ranging from a few square kilometers to entire continents, satellite images offer a global perspective, crucial for monitoring extensive phenomena and long-term changes.

---

[1] https://sentinel.esa.int/web/sentinel/missions.

Spatial resolution varies by sensor and platform. High-resolution satellites unveil details down to meters, while lower-resolution ones provide a broader view of Earth's surface. Temporal resolution hinges on satellite revisit periods, spanning days to weeks or months based on specifics. Atmospheric conditions like cloud cover and haze influence image quality, although some sensors counter these effects, while others, like radar sensors, remain unaffected.

Aerial images are captured from platforms that are closer to the Earth's surface, such as airplanes, helicopters, or drones [33]. These platforms are equipped with cameras or other sensors, allowing for the acquisition of images at specific locations and altitudes. Aerial images offer localized coverage and can be acquired over targeted areas of interest. They are particularly useful for capturing detailed imagery of specific regions, such as cities, construction sites, or natural landscapes. Aerial images generally have higher spatial resolution compared to satellite images. They can capture fine details, objects, and features with greater clarity and precision. The spatial resolution of aerial images can range from centimeters to a few meters, depending on the sensor and flight parameters. Aerial images can be acquired on-demand, allowing for more frequent revisit times compared to satellites. The temporal resolution of aerial images depends on factors like flight scheduling and availability of aircraft or drones. Aerial images are less affected by atmospheric conditions compared to satellite images. Being closer to the Earth's surface, they are captured under relatively clearer atmospheric conditions, resulting in improved image quality and reduced atmospheric interference.

Both satellite and aerial images have their advantages and are used in various applications [31]. Satellite images provide a global perspective and long-term monitoring capabilities, while aerial images offer higher spatial resolution and localized coverage for detailed analysis of specific areas of interest. The choice between satellite and aerial imagery depends on the specific requirements of the application, the desired level of detail, and the availability of data.

## 3   Definition of Semantic Segmentation

Semantic segmentation tasks focus on labeling each pixel of an image with a corresponding class of what the pixel represents. The goal is to produce a dense pixel-wise segmentation map of an image, where specific class is assigned to each pixel. The tasks of image semantic segmentation aim at the fine-grained identification of objects in an image. In contrast to object detection, which aims at coarser localization of the detected objects. Recently, more sophisticated extensions of the semantic segmentation task, referred as instance segmentation [15] and panoptic segmentation [20] have emerged. Instance segmentation takes into account different semantic types and focuses on delineating multiple objects present in an image. On the basis of instance segmentation, panoptic segmentation needs to detect and segment all objects in the image, including the background.

Semantic segmentation of remote sensing images is a fundamental task in the field of remote sensing and computer vision. The goal is to partition the image into meaningful regions, enabling detailed analysis and understanding of the Earth's surface. Figure 1 illustrates an example of semantic segmentation of remote sensing images in the context of buildings extraction.

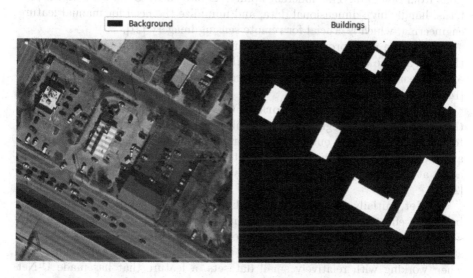

**Fig. 1.** Semantic segmentation of buildings in remote sensing imagery: A sample image (left) and its corresponding output (right) displaying prediction overlays. This example is sourced from the Massachusetts Buildings dataset [25].

With the continuous advancements in semantic segmentation techniques, they have found applications in addressing diverse and data-rich remote sensing problems [31]. These problems often involve complex and high-dimensional datasets, such as aerial and satellite images, which require accurate and detailed analysis. The semantic segmentation of remote sensing images plays an important role in many applications [31].

## 4    Methods for Semantic Segmentation

The task of semantic segmentation in remote sensing images has its unique challenges, due to the high resolution, complex spatial structures, diverse object scales, and the huge amounts of data.

Initially, traditional machine learning methods were the go-to solutions for this task, primarily grouped into two categories: pixel-based methods and region-based methods [46]. Both, pixel-based and region-based methods, relied heavily on handcrafted features and manual or heuristic threshold selection. Although sometimes effective, these methods often struggled with complex images with varying lighting conditions, textures, and scales. As a result, their performance could be inconsistent and their application limited compared to modern deep learning-based approaches [47].

The advent of deep learning brought forth a paradigm shift in semantic segmentation of remote sensing imagery. Deep learning models, unlike traditional machine learning techniques, can automatically learn hierarchical representations from raw data. This inherent ability enables them to detect complex patterns, handle high-dimensional data, and minimize the need for manual feature engineering, which is crucial for remote sensing imagery [21].

One of the most prominent deep learning models applied in the context of this task are the **Convolutional Neural Networks** (CNNs). CNNs are a class of deep learning models that excel at processing grid-like data, such as images [14].

**Fully Convolutional Networks** (FCNs) are an important development in the field of semantic segmentation. Unlike traditional CNNs that are confined to fixed-size inputs and outputs, FCNs are designed to handle inputs of any size and produce corresponding spatial outputs. This quality makes them particularly adept at pixel-level prediction tasks, a key requirement in semantic segmentation [22].

**U-Net**, initially designed for biomedical image segmentation, has proven to be highly effective in the semantic segmentation of remote sensing images as well. A unique mark of the U-Net architecture is its symmetric encoder-decoder structure. Its structure delivers detailed and accurate segmentation maps, even when working with relatively small datasets, a feature that has made U-Net particularly popular for tasks requiring precise localization [13].

**Multi-Scale Contextual Models**, represented by notable architectures such as DeepLabv3 [4] and Pyramid Scene Parsing Network (PSPNet) [48], introduce a novel approach to semantic segmentation that seeks to incorporate context information at varying scales. This is particularly beneficial in remote sensing image analysis where objects of interest often appear at different scales and densities.

**Attention-based** are pivotal in semantic segmentation of remote sensing images due to their capacity to allocate computational focus selectively [12]. They employ an attention mechanism, which uses attention maps, learned during the training process, to assign varying weights to different regions in the feature maps. This capability is particularly crucial for remote sensing images where some regions can be more relevant than others depending on the task at hand. The weights assigned by the attention mechanism help to amplify the influence of important regions and suppress the less important ones in the subsequent layers of the model, enhancing the ability to distinguish different land cover classes or physical objects within the imagery.

**The Masked-attention Mask Transformer** (Mask2Former) architecture falls in this category [6]. It is a novel architecture proficient in managing various image segmentation tasks such as panoptic, instance, or semantic segmentation. The design is built upon a simple meta architecture, which includes a backbone feature extractor, a pixel decoder, and a Transformer decoder. A main feature is the incorporation of masked attention within the Transformer decoder.

The evaluation of the semantic segmentation models in the context of remote sensing images is primarily done using *Pixel Accuracy, Mean Accuracy, Intersection over Union (IoU)* and *F1 Score (Dice Coefficient)*. Further details on the definition of the various metrics can be found in [31].

## 5   Datasets for Semantic Segmentation of Remote Sensing Images

There are numerous datasets specifically designed for semantic segmentation of remote sensing images that have been widely used for training and evaluating algorithms. Table 1 presents a summary of these datasets, including their source, type, number of images, image size, spatial resolution, format and provided bands. The datasets are with different spatial resolutions and sizes. The number of semantic labels in these datasets ranges from 2 to 20.

The datasets in the table exhibit a wide range of sources. While the majority of the datasets consist of aerial RGB images, there are also datasets that include multi-spectral data from specific satellite missions such as Sentinel-1, Sentinel-2, and Gaofen-2 [5]. Additionally, some datasets are sourced from platforms like Google Earth. It is worth noting that a significant portion of the datasets are obtained using unmanned aerial vehicles (UAVs), specifically drones, which provide high-resolution imagery for various applications.

Several datasets are specifically designed to support multi-class semantic segmentation tasks for accurate land cover mapping, such as SEN12MS [39], SemCity Toulouse [37], Christchurch Aerial Semantic Dataset (CASD) [2], DFC2022 [17], DLRSD [40] and Dubai's Satellite Imagery Dataset [28]. The LandCover.ai dataset, also known as Land Cover from Aerial Imagery, is specifically designed for the automatic mapping of buildings, woodlands, water, and roads from aerial images [3]. It contains a selection of aerial images taken over the area of Poland. The Inria Aerial Image Labeling dataset focuses on the task of semantic segmentation of aerial imagery [24] by providing ground truth data for two semantic classes: building and not building.

The Massachusetts Roads dataset consists of 1171 aerial images of the state of Massachusetts [25]. The target maps for the dataset are created by converting road centerlines obtained from the OpenStreetMap project into raster format. The labels are generated without any smoothing, using a line thickness of 7 pixels. The Massachusetts Buildings Dataset is composed of 151 aerial images capturing the Boston area [25]. The target maps for this dataset are generated by converting building footprints obtained from the OpenStreetMap project into raster format. GTA-V [49] is synthetic dataset for remote sensing image segmentation tailored for building extraction.

The Potsdam dataset [38] and Vaihingen dataset [38] are for urban semantic segmentation. These datasets are used in the 2D Semantic Labeling Contest. The datasets encompass five foreground classes: impervious surface, building, low vegetation, tree, car and one background class referred as clutter. The masks for these datasets are 3-channel geotiffs with unique RGB values for each class. The

Remote Sensing Land-Cover dataset for Domain Adaptive Semantic Segmentation (LoveDA) [45] encompasses two scenes (urban and rural) with significant challenges arise from the presence of multi-scale objects, intricate background elements, and uneven class distributions within the dataset. The Gaofen Image Dataset (GID-15) for semantic segmentation [42] contains 150 satellite images. The images are taken by the Gaofen-2 (GF-2) satellite over 60 cities in China. The images are organized into 15 semantic categories. The DeepGlobe Land Cover Classification Challenge dataset is designed for semantic segmentation tasks [9]. It includes high-resolution sub-meter satellite imagery for classifying land cover categories.

The Cloud Cover Segmentation Dataset was created through a crowdsourcing competition and subsequently validated by a team of expert annotators [35]. This dataset comprises Sentinel-2 satellite imagery along with corresponding cloud labels stored as GeoTiffs. The 95-Cloud dataset is an expansion of the previously released 38-Cloud dataset for cloud detection [27]. This binary classification allows for the precise identification and separation of cloudy areas within the imagery.

FloodNet, as described in [36], offers high-resolution UAS imagery with detailed semantic annotations specifically focusing on damage assessment after Hurricane Harvey. The dataset is captured using DJI Mavic Pro quadcopters, providing valuable information for flood damage analysis. The semantic annotations in FloodNet offer precise labeling for various classes, enabling accurate assessment of flood-related damages. ETCI2021 Flood Detection dataset [30] contains data from the flood event detection contest, organized by the NASA Interagency Implementation and Advanced Concepts Team. The primary objective of the dataset is to foster innovation in the detection of flood events and water bodies.

RIT-18 [18] dataset contains very-high resolution multispectral imagery collected by an unmanned aircraft system. The primary use of this dataset is for evaluating semantic segmentation frameworks designed for non-RGB remote sensing imagery. Several datasets, such as UAVid [23], Aeroscapes [32], DroneDeploy [11] and Semantic Drone Dataset [43] are specifically designed to enhance semantic understanding of urban scenes. These semantic segmentation datasets are with fine-resolution images obtained using UAVs.

## 6    Applications Enabled by Semantic Segmentation in Remote Sensing

With the ever-expanding number of aerial and satellite images, coupled with the constant development of modern deep-learning techniques, the application of semantic segmentation of aerial images takes significant contribution to wide range of applications, like urban planning, land cover mapping, disaster management, environmental monitoring, and precision agriculture.

Semantic segmentation of remote sensing images, can help urban planners to gain insights into various aspects of urban environments, including the identification of buildings, roads, vegetation, water bodies, and other infrastructure

**Table 1.** Properties of the different semantic segmentation datasets. The table contains information for the source, type, number of images, image size, spatial resolution, format and provided bands for the different datasets.

| Name | Source | #Images | Image Size | Spatial Res. | #Labels | Format | Bands |
|---|---|---|---|---|---|---|---|
| LandCover.ai [3] | Aerial | 10674 | 512 × 512 | 0.25–0.5 m | 5 | geo tif | RGB |
| Inria [24] | Aerial | 360 | 5000 × 5000 | 0.3 m | 2 | geo tif | RGB |
| Massachusetts Roads [25] | Aerial | 1171 | 1500 × 1500 | 1 m | 2 | tif | RGB |
| Massachusetts Buildings [25] | Aerial | 151 | 1500 × 1500 | 1 m | 2 | tif, png | RGB |
| Vaihingen [38] | Aerial | 33 | 2494 × 2064 | 0.09 m | 6 | geo tif | RG, NIR |
| Potsdam [38] | Aerial | 38 | 6000 × 6000 | 0.05 m | 6 | geo tif | RGB, NIR |
| LoveDA [45] | Google Earth | 5987 | 1024 × 1024 | 0.03 m | 7 | png | RGB |
| GID-15 [42] | Gaofen-2 | 150 | 6800 × 7200 | 3 m | 15 | tif | RGB |
| DeepGlobe Land Cover [9] | DigitalGlobe Vivid+ | 803 | 2448 × 2448 | 0.5 m | 7 | png, jpg | RGB |
| UAVid [23] | UAV | 420 | 4096 × 2160 | n/a | 8 | png | RGB |
| Cloud Cover Segmentation [35] | Sentinel-2 | 22728 | 512 × 512 | 10 m | 2 | geo tif | MSI |
| 95-Cloud [27] | Landsat 8 | 43902 | 384 × 384 | 30 m | 2 | png | RGB, NIR |
| ETCI2021 Flood Detection [30] | Sentinel-1 | 66810 | 256 × 256 | 5–20 m | 2 | png | SAR |
| FloodNet [36] | UAV | 2343 | 4000 × 3000 | 0.015 m | 10 | jpg, png | RGB |
| SEN12MS [39] | Sentinel-1/2, MODIS | 180662 | 256 × 256 | 10 m | 33 | tif | SAR, MSI |
| CASD [2] | Aerial | 4 | 4800 × 3600 | 0.1 m | 4 | tif | RGB |
| SemCity Toulouse [37] | Worldview-II | 16 | 3504 × 3452 | 2 m | 7 | geo tif | MSI |
| DFC2022 [17] | Aerial | 3981 | 2000 × 2000 | 0.5 m | 16 | geo tif | RGB |
| DLRSD [40] | Aerial | 2100 | 256 × 256 | 0.305 m | 17 | tif, png | RGB |
| Dubai's Satellite Imagery Dataset [28] | Aerial | 72 | different sizes | n/a | 6 | jpg, png | RGB |
| GTA-V-SID [49] | Synthetic | 121 | 500 × 500 | 1 m | 2 | png | RGB |
| RIT-18 [18] | Synthetic | 3 | different sizes | 0.047 m | 19 | mat | MSI |
| Aeroscapes [32] | UAV | 3269 | 1280 × 720 | n/a | 12 | jpg, png | RGB |
| DroneDeploy [11] | UAV | 55 | 11084 × 12326 | 0.1 m | 7 | tif, png | RGB |
| Semantic Drone Dataset [43] | UAV | 600 | 6000 × 4000 | n/a | 20 | png | RGB |

elements [3,24]. Further more, it can helps in understanding the spatial distribution and patterns of the different land cover types, which is crucial for urban planning tasks such as infrastructure development and transportation planning. The semantic segmentation of remote sensing images can support the analysis of urban growth and change over time. The study presented in [41] introduces an approach that involves comparing segmented images taken at various time intervals. This allows urban planners to analyze the dynamics of urban development, monitor changes in land use, and evaluate the impact of urban planning policies and interventions.

During natural disasters such as floods, earthquakes, or wildfires, remote sensing images can be used to monitor the affected areas and detect changes over time [7,16], [?]. In [16], deep learning techniques including PSPNet, DeepLabV3, and U-Net are employed on the FloodNet dataset [36] to detect floods. The focus of the study is on identifying flooded roads and buildings, as well as distinguishing between natural water and flooded water.

By leveraging the spatial and temporal resolution of remote sensing images, semantic segmentation can also facilitate the monitoring and management of post-disaster recovery and reconstruction activities. In [7], the authors present an improved Swin transformer for semantic segmentation of post-earthquake dense buildings in urban areas. The method is used to identify damaged buildings, allowing emergency response teams to prioritize rescue and recovery efforts.

One of the primary applications of semantic segmentation in environmental monitoring is the detection and monitoring of deforestation. By segmenting satellite or aerial images, semantic segmentation algorithms can identify forested areas and distinguish them from cleared or degraded regions. In [1], the authors propose a specialized variant of the DeepLabv3+ architecture called DeepLab Change Detection (DLCD) for detecting deforestation in areas of significant forest loss.

Climate monitoring is another important aspect of environmental monitoring that benefits from semantic segmentation. By analyzing remote sensing images, semantic segmentation can identify and classify different climate-related features, such as clouds, aerosols, atmospheric conditions, and surface temperature variations [27,35]. This information aids in climate modeling, weather forecasting, and understanding the dynamics of climate change.

By segmenting remote sensing images, semantic segmentation can identify water bodies, such as rivers, lakes, and reservoirs, as well as detect changes in water levels and quality. Further more, Semantic segmentation aids in mapping and monitoring wetlands, coastal areas, and other ecologically sensitive water-based ecosystems. In [29], the authors utilize two semantic segmentation methods, namely DeepLabv3+ and SegNet, for the detection of water bodies. The objective of this detection is to estimate water levels and monitor temporal fluctuations in water levels.

Important aspect of precision agriculture is crop health monitoring. Semantic segmentation can detect and classify vegetation health indicators, such as areas affected by pests, diseases, nutrient deficiencies, or water stress. For example, in

[8] a high-resolution aerial imagery and UNet with a convolutional LSTM is used to accurately detect regions of the field showing nutrient deficiency stress. This approach minimizes resource wastage and optimizes crop health and yield. Water management is another critical component of precision agriculture. By analyzing remote sensing images, semantic segmentation can map and monitor soil moisture levels, water stress, and irrigation efficiency across agricultural fields. Furthermore, semantic segmentation aids in weed detection and management as in [26] where semantic segmentation is applied in two stages using UNet architecture. This approach promotes sustainable weed control practices and reduces the development of herbicide resistance.

## 7   Conclusion

In conclusion, semantic segmentation of remote sensing images is a powerful technique that enables the accurate classification and labeling of individual pixels or regions within an image, based on their semantic meaning. It plays a crucial role in various applications across different domains. The paper provides a comprehensive overview of semantic segmentation, starting with its definition and the underlying methods used for image analysis and classification. It discusses the importance of high-quality datasets for training and evaluation purposes, highlighting several notable datasets available for semantic segmentation of remote sensing images.

Furthermore, the paper explores the wide range of applications where semantic segmentation is utilized. It covers areas such as urban planning, disaster management, environmental monitoring, precision agriculture, deforestation, climate analysis, water management, and more. Each application benefits from the accurate and detailed understanding of land cover and object classification provided by semantic segmentation. The diverse datasets described in the paper, ranging from aerial and satellite imagery to UAV and synthetic data, reflect the breadth of sources used for remote sensing image analysis. The inclusion of different spectral bands and data formats demonstrates the flexibility and adaptability of semantic segmentation techniques.

**Acknowledgements.** This work was partially financed by the Faculty of Computer Science and Engineering at the Ss. Cyril and Methodius University in Skopje.

## References

1. de Andrade, R.B., Mota, G.L.A., da Costa, G.A.O.P.: Deforestation detection in the amazon using deeplabv3+ semantic segmentation model variants. Remote Sens. **14**(19) (2022). https://doi.org/10.3390/rs14194694, https://www.mdpi.com/2072-4292/14/19/4694
2. Audebert, N., Saux, B.L., Lefèvre, S.: Segment-before-detect: vehicle detection and classification through semantic segmentation of aerial images. Remote. Sens. **9**, 368 (2017)

3. Boguszewski, A., Batorski, D., Ziemba-Jankowska, N., Dziedzic, T., Zambrzycka, A.: LandCover.ai: dataset for automatic mapping of buildings, woodlands, water and roads from aerial imagery. In: Proceedings of the IEEE/CVF Conference on Computer Vision and Pattern Recognition (CVPR) Workshops, pp. 1102–1110 (2021)

4. Chen, L.C., Papandreou, G., Kokkinos, I., Murphy, K., Yuille, A.L.: DeepLab: semantic image segmentation with deep convolutional nets, Atrous convolution, and fully connected CRFs. IEEE Trans. Pattern Anal. Mach. Intell. **40**(4), 834–848 (2017)

5. Chen, L., et al.: An introduction to the Chinese high-resolution earth observation system: Gaofen-1 7 civilian satellites. J. Remote Sens. **2022** (2022)

6. Cheng, B., Choudhuri, A., Misra, I., Kirillov, A., Girdhar, R., Schwing, A.G.: Mask2former for video instance segmentation. arXiv preprint arXiv:2112.10764 (2021)

7. Cui, L., Jing, X., Wang, Y., Huan, Y., Xu, Y., Zhang, Q.: Improved swin transformer-based semantic segmentation of postearthquake dense buildings in urban areas using remote sensing images. IEEE J. Sel. Top. Appl. Earth Observ. Remote Sens. **16**, 369–385 (2023). https://doi.org/10.1109/JSTARS.2022.3225150

8. Dadsetan, S., Rose, G.L., Hovakimyan, N., Hobbs, J.: Detection and prediction of nutrient deficiency stress using longitudinal aerial imagery. In: AAAI Conference on Artificial Intelligence (2020)

9. Demir, I., et al.: DeepGlobe 2018: a challenge to parse the earth through satellite images. In: The IEEE Conference on Computer Vision and Pattern Recognition (CVPR) Workshops (2018)

10. Dimitrovski, I., Kitanovski, I., Kocev, D., Simidjievski, N.: Current trends in deep learning for earth observation: an open-source benchmark arena for image classification. ISPRS J. Photogramm. Remote. Sens. **197**, 18–35 (2023)

11. DroneDeploy: Dronedeploy machine learning segmentation benchmark (2019). https://github.com/dronedeploy/dd-ml-segmentation-benchmark. Accessed 7 June 2023

12. Fu, J., et al.: Dual attention network for scene segmentation. In: 2019 IEEE/CVF Conference on Computer Vision and Pattern Recognition (CVPR), pp. 3141–3149 (2019)

13. Gu, X., Li, S., Ren, S., Zheng, H., Fan, C., Xu, H.: Adaptive enhanced swin transformer with u-net for remote sensing image segmentation. Comput. Electr. Eng. **102**, 108223 (2022)

14. Guo, Y., Liu, Y., Georgiou, T., Lew, M.S.: A review of semantic segmentation using deep neural networks. Int. J. Multimed. Inf. Retrieval **7**, 87–93 (2018)

15. Hafiz, A.M., Bhat, G.: A survey on instance segmentation: state of the art. Int. J. Multimed. Inf. Retrieval **9** (2020)

16. Hernández, D., Cecilia, J.M., Cano, J.C., Calafate, C.T.: Flood detection using real-time image segmentation from unmanned aerial vehicles on edge-computing platform. Remote Sens. **14**(1) (2022)

17. Hänsch, R., et al.: Data fusion contest 2022 (DFC2022) (2022). https://dx.doi.org/10.21227/rjv6-f516

18. Kemker, R., Salvaggio, C., Kanan, C.: Algorithms for semantic segmentation of multispectral remote sensing imagery using deep learning. ISPRS J. Photogram. Remote Sens. **145**, 60–77 (2018)

19. King, M., Herring, D.: Satellites | research (atmospheric science). In: Holton, J.R. (ed.) Encyclopedia of Atmospheric Sciences, pp. 2038–2047. Academic Press, Oxford (2003)

20. Kirillov, A., He, K., Girshick, R., Rother, C., Dollar, P.: Panoptic segmentation. In: Proceedings of the IEEE/CVF Conference on Computer Vision and Pattern Recognition (CVPR) (2019)
21. Lateef, F., Ruichek, Y.: Survey on semantic segmentation using deep learning techniques. Neurocomputing **338**, 321–348 (2019)
22. Long, J., Shelhamer, E., Darrell, T.: Fully convolutional networks for semantic segmentation. In: Proceedings of the IEEE Conference on Computer Vision and Pattern Recognition, pp. 3431–3440 (2015)
23. Lyu, Y., Vosselman, G., Xia, G.S., Yilmaz, A., Yang, M.Y.: UAVid: a semantic segmentation dataset for UAV imagery. ISPRS J. Photogramm. Remote. Sens. **165**, 108–119 (2020)
24. Maggiori, E., Tarabalka, Y., Charpiat, G., Alliez, P.: Can semantic labeling methods generalize to any city? The INRIA aerial image labeling benchmark. In: IEEE International Geoscience and Remote Sensing Symposium (IGARSS) (2017)
25. Mnih, V.: Machine learning for aerial image labeling. Ph.D. thesis, University of Toronto (2013)
26. Moazzam, S.I., Khan, U.S., Qureshi, W.S., Nawaz, T., Kunwar, F.: Towards automated weed detection through two-stage semantic segmentation of tobacco and weed pixels in aerial imagery. Smart Agric. Technol. **4**, 100142 (2023)
27. Mohajerani, S., Saeedi, P.: Cloud-net+: a cloud segmentation CNN for landsat 8 remote sensing imagery optimized with filtered Jaccard loss function, vol. 2001.08768 (2020)
28. Mohammed Bin Rashid Space Center: Semantic segmentation dataset (2020). https://humansintheloop.org/resources/datasets/semantic-segmentation-dataset-2/. Accessed 7 June 2023
29. Muhadi, N.A., Abdullah, A.F., Bejo, S.K., Mahadi, M.R., Mijic, A.: Deep learning semantic segmentation for water level estimation using surveillance camera. Appl. Sci. **11**(20) (2021). https://doi.org/10.3390/app11209691, https://www.mdpi.com/2076-3417/11/20/9691
30. NASA Interagency Implementation and Advanced Concepts Team: ETCI 2021 competition on flood detection (2021). https://nasa-impact.github.io/etci2021/. Accessed 7 June 2023
31. Neupane, B., Horanont, T., Aryal, J.: Deep learning-based semantic segmentation of urban features in satellite images: a review and meta-analysis. Remote Sens. **13**(4), 808 (2021)
32. Nigam, I., Huang, C., Ramanan, D.: Ensemble knowledge transfer for semantic segmentation. In: 2018 IEEE Winter Conference on Applications of Computer Vision (WACV), pp. 1499–1508 (2018)
33. Osco, L.P., et al.: A review on deep learning in UAV remote sensing. Int. J. Appl. Earth Obs. Geoinf. **102**, 102456 (2021)
34. Pearlman, J., Barry, P., Segal, C., Shepanski, J., Beiso, D., Carman, S.: Hyperion, a space-based imaging spectrometer. IEEE Trans. Geosci. Remote Sens. **41**(6), 1160–1173 (2003)
35. Radiant Earth Foundation: Sentinel-2 cloud cover segmentation dataset (version 1) (2022). https://doi.org/10.34911/rdnt.hfq6m7. Accessed 7 June 2023
36. Rahnemoonfar, M., Chowdhury, T., Sarkar, A., Varshney, D., Yari, M., Murphy, R.R.: FloodNet: a high resolution aerial imagery dataset for post flood scene understanding. IEEE Access **9**, 89644–89654 (2021)
37. Roscher, R., Volpi, M., Mallet, C., Drees, L., Wegner, J.D.: Semcity toulouse: a benchmark for building instance segmentation in satellite images. ISPRS Ann. Photogram. Remote Sens. Spat. Inf. Sci. **V-5-2020**, 109–116 (2020)

38. Rottensteiner, F., et al.: The ISPRS benchmark on urban object classification and 3D building reconstruction. ISPRS Ann. Photogram. Remote Sens. Spat. Inf. Sci. **I-3** (2012)
39. Schmitt, M., Wu, Y.L.: Remote sensing image classification with the sen12ms dataset. In: ISPRS Ann. Photogram. Remote Sens. Spat. Inf. Sci. **V-2-2021**, 101–106 (2021)
40. Shao, Z., Yang, K., Zhou, W.: Performance evaluation of single-label and multi-label remote sensing image retrieval using a dense labeling dataset. Remote Sens. **10**(6), 964 (2018)
41. Toker, A., et al.: DynamicEarthNet: daily multi-spectral satellite dataset for semantic change segmentation. In: Proceedings of the IEEE/CVF Conference on Computer Vision and Pattern Recognition (CVPR), pp. 21158–21167 (2022)
42. Tong, X.Y., et al.: Land-cover classification with high-resolution remote sensing images using transferable deep models. Remote Sens. Environ. **237**, 111322 (2020)
43. TU Graz, Graz University of Technology : Semantic drone dataset (2020). http://dronedataset.icg.tugraz.at. Accessed 7 June 2023
44. Tupin, F., Inglada, J., Nicolas, J.M.: Remote Sensing Imagery. Wiley, New York (2014)
45. Wang, J., Zheng, Z., Ma, A., Lu, X., Zhong, Y.: LoveDA: a remote sensing land-cover dataset for domain adaptive semantic segmentation. In: Vanschoren, J., Yeung, S. (eds.) Proceedings of the Neural Information Processing Systems Track on Datasets and Benchmarks, vol. 1. Curran Associates, Inc. (2021)
46. Weinland, D., Ronfard, R., Boyer, E.: A survey of vision-based methods for action representation, segmentation and recognition. Comput. Vis. Image Underst. **115**(2), 224–241 (2011)
47. Yuan, X., Shi, J., Gu, L.: A review of deep learning methods for semantic segmentation of remote sensing imagery. Expert Syst. Appl. **169**, 114417 (2021)
48. Zhao, H., Shi, J., Qi, X., Wang, X., Jia, J.: Pyramid scene parsing network. In: Proceedings of the IEEE Conference on Computer Vision and Pattern Recognition, pp. 2881–2890 (2017)
49. Zou, Z., Shi, T., Li, W., Zhang, Z., Shi, Z.: Do game data generalize well for remote sensing image segmentation? Remote Sens. **12**(2), 275 (2020)

# Enhancing Knee Meniscus Damage Prediction from MRI Images with Machine Learning and Deep Learning Techniques

Martin Kostadinov[1], Petre Lameski[1], Andrea Kulakov[1], Ivan Miguel Pires[2,3], Paulo Jorge Coelho[4,5], and Eftim Zdravevski[1]([✉])

[1] Faculty of Computer Science and Engineering, University of Sts. Cyril and Methodius in Skopje, Macedonia, Ruger Boskovik 16, Skopje 1000, Macedonia
eftim.zdravevski@finki.ukim.mk
[2] Instituto de Telecomunicações, Aveiro, Portugal
[3] Escola Superior de Gestão e Tecnologia, Politécnico de Santarém, Santarém, Portugal
[4] Institute for Systems Engineering and Computers at Coimbra (INESC Coimbra), DEEC, Coimbra, Portugal
[5] School of Technology and Management, Polytechnic of Leiria, Leiria, Portugal

**Abstract.** This paper investigates the application of machine learning and deep learning models to predict knee meniscus damage from magnetic resonance imaging (MRI) scans. We utilized the MRNet dataset, and processed it with different approaches, using a one-dimensional grayscale, RGB, and segmented images, complemented with features extracted using Histogram of Oriented Gradients (HOG) and Scale-Invariant Feature Transform (SIFT) techniques. Our objective was to evaluate whether a DL model could match or exceed the diagnostic performance of clinical experts such as general radiologists and orthopedic surgeons. Our findings demonstrate that our ML and DL models can predict meniscal tears with comparable accuracy to that of general medical doctors. This suggests that ML and DL models have potential to deliver rapid preliminary results post-MRI exams and augment the quality of MRI diagnoses, particularly in settings lacking specialist radiologists. Thus, integrating ML and DL models into clinical practice could enhance the quality and consistency of MRI interpretation for knee meniscus damage.

**Keywords:** MRI images · machine learning · deep learning · transformer architecture · disease prediction · knee meniscus prediction

M. Mihova and M. Jovanov (Eds.): ICT Innovations 2023, CCIS 1991, pp. 141–155, 2024.
https://doi.org/10.1007/978-3-031-54321-0_10

# 1   Introduction

The progress made in medical imaging and deep learning (DL) has made it possible to efficiently analyze vast image databases. This includes the analysis of images from computed tomography (CT), magnetic resonance imaging (MRI), and X-ray sources [1]. With these techniques, scientists are able to utilize deep neural networks to obtain image representations and conduct further examinations using clustering, visualization via dimensionality reduction, and class activation mapping. Such medical approaches could be beneficial for the elderly people, or people going through recovery process [2,3].

In the beginning, radiomics and deep learning methods were concentrated on binary recognition tasks like identifying the existence or non-existence of a single anomaly, and distinguishing between benign and malignant [4]. Contemporary algorithms strive to encompass meaningful multiclass categorization of identified irregularities, such as the classification and malignancy assessment of tumors. Contemporary methodologies further forecast the success rates of treatment, the completeness of surgical removal, and the risk of relapse [4].

Situated within the knee joint, the meniscus is a crescent-shaped cartilage component, serving as a cushion between the thigh bone (femur) and the shinbone (tibia) [5]. Meniscus knee injuries commonly occur due to twisting or rotating the knee forcefully while bearing weight [6]. These injuries can lead to a range of problems and complications [7,8]. When associated with trauma, due to sudden, forceful movements or direct blows to the knee - often encountered during sports activities or accidents, can result in meniscus tears [9]. When associated with age-related degeneration, the meniscus can become weaker and more prone to tears or damage, even from relatively minor movements [10].

The major problems associated with meniscus knee injuries are pain and swelling, that can vary depending on the extent and location of the tear. For athletes, this can significantly impact an individual's ability to participate in sports or physical activities, hindering their performance and potentially leading to a decline in overall fitness [11]. Also, the limited range of motion as a torn meniscus can restrict the normal movement of the knee joint, making it difficult to fully bend or straighten the leg. This situation may lead to instability in the knee joint. This instability can cause a feeling of "giving way" or the knee buckling during activities. Furthermore, the locking or catching sensations, as a torn meniscus can result in fragments of the cartilage moving into the joint space [12]. This can cause the knee to lock or catch during movement, leading to discomfort and limited mobility. If a meniscus injury is not treated promptly or properly, it can contribute to the development of osteoarthritis in the knee joint over time. The loss of the meniscus's protective cushioning can lead to increased friction and wear on the joint surfaces. It is worth noting that the severity and symptoms of meniscus knee injuries can vary widely [13]. Treatment options depend on factors such as the location, size, and type of tear, as well as the individual's age, activity level, and overall health. Mild cases may be managed with conservative measures like rest, ice, compression, and physical therapy, while more severe tears may require surgical intervention, such as arthroscopic repair or partial meniscectomy [14].

The diagnosis, however, is made in most cases through magnetic resonance imaging (MRI) and predicting knee meniscus damage from MRI images using machine learning (ML) and deep learning techniques (DL) can be a valuable approach to assist medical professionals in diagnosing and treating patients [15, 16].

This study contributes to demonstrate that ML and DL models can predict meniscal tears with similar accuracy to that of physicians, presenting the potential to enhance the quality of MRI diagnoses and provide quick preliminary results after MRI tests, especially in situations without specialized radiologists.

Section 2 presents an overview of some of the existing approaches in the literature. Then, in Sect. 3 are described the methods for processing the dataset and the characteristics of the dataset. Afterwards, in Sect. 4 the implementations and challenges are presented. Next, in Sect. 5 are provided the results obtained, and finally in Sect. 6, some conclusions of the paper and discussion of some ideas for future work are presented.

## 2 Related Work

In recent years, there has been growing interest in utilizing machine learning and deep learning techniques to enhance the prediction of knee meniscus damage from MRI images. Several studies have explored various approaches and methodologies in this domain.

Regarding Meniscal Tear Classification, some previous studies have focused on classifying meniscal tears into different categories based on their location, size, and shape. Various classification schemes, such as the International Society of Arthroscopy, Knee Surgery, and Orthopedic Sports Medicine (ISAKOS) classification, have been proposed to standardize tear categorization [17]. These studies provide a foundation for accurately characterizing meniscal tears and assessing their severity.

For feature Extraction Techniques, researchers have explored different extraction methods to capture relevant information from MRI images for meniscus damage prediction [18]. These techniques include shape analysis, texture analysis, intensity histograms, and local binary patterns [19]. By extracting discriminative features, these studies aim to enhance the predictive capability of machine learning models.

In Machine Learning Approaches, studies have employed machine learning algorithms for knee meniscus damage prediction. Support Vector Machines (SVM), Random Forests, and Naive Bayes classifiers have been utilized to classify meniscal tears based on extracted features [20]. The authors of [20] sought to develop and test an algorithm, using 1123 MR images of the knee, for identifying and describing a meniscus tear. This goal was fulfilled by dividing the principal task into three distinct phases: commencing with the identification of the positions of both horns, followed by the detection of a tear and subsequent determination of its orientation. Employing CNNs, the outcomes yielded an AUC of 0.92 for horn localization, 0.94 for tear detection, and 0.83 for tear orientation determination.

More recently with Deep Learning Models, Convolutional Neural Networks (CNNs) have been extensively used for knee meniscus damage prediction [21]. CNNs have showcased exceptional effectiveness in tasks involving image classification, achieved through their inherent capacity to autonomously glean hierarchical features from unprocessed MRI images. Studies have utilized pre-trained CNN architectures, such as VGGNet and ResNet, and fine-tuned them for meniscus damage classification [22].

The methodology introduced in [23] employed a bespoke ResNet-14 architecture, comprising 14 layers, within a CNN framework. This architecture was configured with six distinct directions and utilized techniques like class balancing and data augmentation. The diagnostic outcomes presented in the study suggested that the proposed deep-learning approach could be effectively employed to automatically detect and assess ACL injuries in athletes.

The researchers in [24] utilized a collection of 1370 knee MRI examinations to create MRNet, a convolutional neural network tailored for the categorization of MRI sequences. By amalgamating predictions from three sequences for each examination, they attained an AUC of 0.937 on an internal dataset and an AUC of 0.824 on an external dataset. These outcomes support the claim that deep learning algorithms can augment the efficiency of clinical specialists in interpreting medical imaging, as does [1].

## 3    Dataset

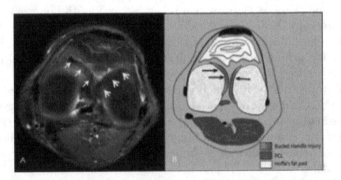

Fig. 1. A) MRI knee example with a tear lesion; B) Anatomical illustration of knee.

### 3.1    Basic Information

The dataset comprises 1,350 MRI panoramic images of the knee joint and corresponding labels for 2-classes, indicating damaged meniscus and undamaged meniscus. Figure 1 presents an example of an injured knee. At the left an MRI image with white arrows indicating the lesion, and at the right, an illustration of the knee segmented by anatomical regions. Dividing the dataset resulted in 1200 images allocated to the training set, accounting for approximately 90%, and 150 images designated for the test set, comprising nearly 10%. The dataset, which can be accessed publicly[1], has been detailed in a previous work [24]. A collection of MRI images sampled from the dataset is showcased in Fig. 2.

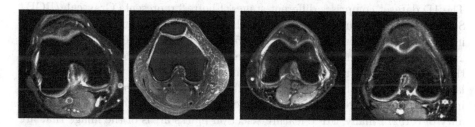

**Fig. 2.** Example of MRI images from dataset.

A variety of datasets or image processing techniques were generated for this study. The grayscale images/flatten dataset refers to a dataset consisting of grayscale images where each image has been flattened into a 1D array.

Grayscale images contain shades of gray from black to white and are represented by a single intensity value per pixel. Similarly, the RGB images/flatten dataset contains RGB (Red, Green, Blue) images that have been flattened into a 1D array [25]. RGB images consist of three color channels, and each pixel is represented by three intensity values (red, green, and blue). The grayscale images can be derived from RGB images. Segmented images/flatten dataset comprises segmented images where objects or regions of interest have been identified and separated from the background. The segmented images have then been flattened into a 1D array for further processing or analysis.

Histogram of Oriented Gradients (HOG) stands as a prevalent technique in computer vision for the extraction of features. [26]. The HOG features dataset contains HOG features extracted from images. HOG features capture local shape and texture information and can be used for tasks like object detection or image classification. Scale-Invariant Feature Transform (SIFT) is another feature extraction technique used in computer vision [27]. Within the SIFT features dataset, one can find a collection of SIFT features derived from images. These SIFT features exhibit resilience against alterations in scale, rotation, and different circumstances, rendering them valuable across a range of computer vision applications. Gabor filters (GF) are a type of linear filter used to analyze the frequency content of an image [28]. The Gabor filter dataset contains image representations obtained by applying Gabor filters. Gabor features capture information about texture and orientation in images.

---

[1] https://stanfordmlgroup.github.io/competitions/mrnet/.

Finally, the mask <225 suggests that a mask has been applied to the images based on a Hounsfield unit threshold. Hounsfield units are used in computed tomography (CT) imaging to quantify the radiodensity of tissues. A mask value of less than 225 is likely applied to isolate soft tissues while excluding bone, as soft tissues typically have Hounsfield unit values ranging from −700 to 225.

### 3.2  1D Dataset

The 1D dataset comprises different stages. Figure 3 represents Grayscale/RGB/ Segmented, and it suggests that for each grayscale, RGB, or segmented image in your dataset, there are a total of 65,536 features. Since the image size is $256 \times 256$ pixels, it implies that each feature corresponds to a specific pixel in the image. Figure 4 represents HOG feature selection that transforms a regular image, applying the HOG technique that extracts local gradient information, into a grayscale image. In this case, the resulting feature vector has a dimensionality of 8,192. Finally, SIFT feature selection marks areas of the image that are invariant during scaling and rotation. It depends on the image and its content. In this case, the approximate number of features extracted is around 98,000.

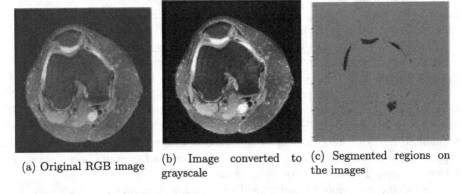

(a) Original RGB image    (b) Image converted to grayscale    (c) Segmented regions on the images

**Fig. 3.** Example of images from the different datasets produced.

## 4  Methods

There are some challenges related to the dataset characteristics and the specific problem you are trying to tackle.

As we have a dataset with many columns (features) but only a few rows (samples), it can lead to overfitting issues. Overfitting occurs when a model learns the noise or specific patterns in the training data too well, making it perform poorly on unseen data. This can be a challenge because with limited samples, the model may struggle to generalize well. To mitigate overfitting, you

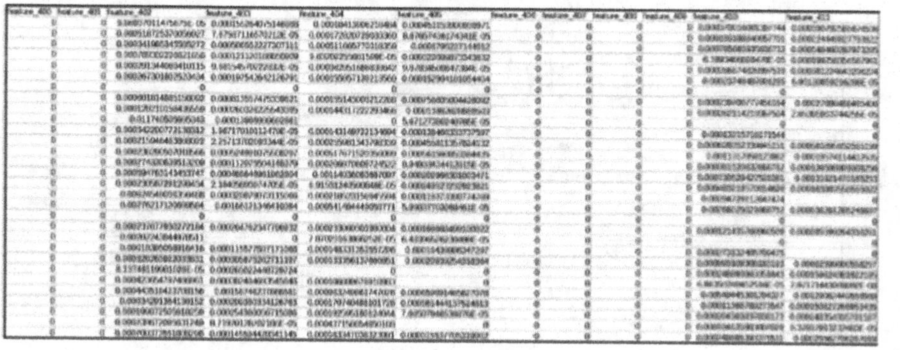

**Fig. 4.** Example of the 1D HOG_features

can consider techniques such as regularization, cross-validation, or reducing the dimensionality of the data.

Principal Component Analysis (PCA) with retention of 95% of the variance of the data distribution. In some cases, retaining 85% of the variance was found to be a better option. PCA serves as a dimensionality reduction method employed to convert high-dimensional data into a reduced-dimensional form, all the while preserving a significant portion of the variance. The choice of retaining 95% or 85% of the variance depends on the specific problem and the trade-off between dimensionality reduction and information preservation. Retaining 95% of the variance will result in a higher-dimensional representation compared to retaining 85%. The decision depends on the requirements of the problem, the available computational resources, and the impact on the achievement in subsequent tasks.

The challenge of learning the "meniscus" that exists only in a small part of the MRI image: If the target or the area of interest you want to learn (the "meniscus") exists only in a small part of the MRI image, it poses a challenge for the learning algorithm to focus on that specific region. This problem is commonly encountered in tasks where the target is sparse or localized. One approach to address this challenge is to use techniques like image segmentation or object detection to identify and extract the region of interest before training the model. By focusing the learning process on the relevant area, you can enhance the model's capability to identify or classify the "meniscus" accurately.

These challenges highlight some common issues encountered in machine learning tasks, such as overfitting, dimensionality reduction, and handling sparse or localized targets.

### 4.1 Machine Learning

Machine learning stands as a domain within artificial intelligence that concentrates on devising algorithms and models, enabling computers to learn from data and subsequently formulate forecasts or choices devoid of explicit programming. This encompasses the process of training a model with a dataset and employing

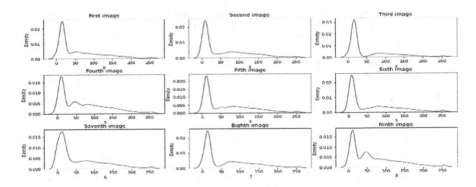

**Fig. 5.** Distribution of the RGB dataset

the trained model to make predictions or judgments concerning novel, unobserved data.

Different machine learning methods were used, including decision tree, logistic regression, K-Nearest Neighbors Classifier, Gaussian Classifier, and XGBoost Classifier.

A decision tree takes on the appearance of a flowchart, wherein every internal node signifies a feature, each branch embodies a decision rule, and each terminal node signifies an outcome. Decision tree classifiers find application in both classification and regression tasks. They partition the feature space into regions and assign a class label to each region.

Logistic regression is a statistical algorithm used for the context of binary classification problems, it establishes a connection between a collection of input features and a binary result through the utilization of a logistic function. It estimates the probability of the outcome being in a particular class based on the input features.

The k-nearest neighbors (KNN) algorithm functions as a non-parametric classification technique. When presented with a new data point, it examines the k nearest training instances within the feature space and assigns the class label based on the majority class among its neighbors. KNN is simple to implement and is often used as a baseline algorithm in classification tasks.

The Gaussian classifier, also known as Gaussian Naive Bayes is an uncomplicated probabilistic classifier grounded in Bayes' theorem with the assumption of independence between features. It models the likelihood of each class using Gaussian distributions and predicts the class with the highest posterior probability.

XGBoost (eXtreme Gradient Boosting) is an ensemble machine learning algorithm that uses gradient boosting framework. It is designed to optimize performance and speed in both classification and regression tasks. XGBoost sequentially trains an ensemble comprises modest predictive models, often in the form of decision trees, and merges their predictions to arrive at the ultimate forecast.

## 4.2   Deep Learning

Deep learning stands as a subset within the realm of machine learning, emphasizing the utilization of artificial neural networks to portray and comprehend intricate patterns and associations within data. These deep learning models consist of numerous tiers of interlinked nodes (neurons) that manipulate and reshape input data, ultimately leading to the formulation of forecasts or judgments.

Neural networks serve as the foundational elements of deep learning models. These networks consist of interconnected units known as neurons, structured into layers. Each neuron accepts inputs, employs an activation function to generate an output, and transfers this output to the subsequent layer. The training of neural networks employs backpropagation, a technique in which the model adapts its internal parameters to minimize the disparity between projected and actual outputs.

Convolutional neural networks (CNNs) represent a specialized variant of neural networks that find widespread application in tasks related to image classification and computer vision. CNNs harness the idea of convolution, wherein filters (small matrices) are employed to extract features from input images. These networks also incorporate pooling layers to diminish spatial dimensions and fully connected layers for classification purposes. CNNs have attained remarkable achievements in the realm of image recognition endeavors.

Finally, Graph Convolutional Networks emerge as a category of deep learning models meticulously tailored for the manipulation of data structured in graphs. Graphs encompass nodes, signifying entities, and edges, symbolizing connections between nodes. GCNs can learn node representations by aggregating information from neighboring nodes, enabling the model to capture local and global graph structures. They have been successfully applied to tasks such as social network analysis, recommendation systems, and molecular chemistry. However, training GCNs can be computationally expensive and memory-intensive, especially for large graphs.

These concepts form the basis of many advanced deep learning techniques and architectures. Deep learning has demonstrated extraordinary accomplishments across diverse domains, encompassing computer vision, natural language processing, speech recognition, and recommendation systems. However, it's important to note that deep learning models often require significant computational resources, and training complex models can be resource-intensive.

## 4.3   ViT: Vision Transformer (Fine-Tune)

The Vision Transformer (ViT) is a deep learning model architecture utilizes the transformer architecture, initially devised for tasks in natural language processing, for applications within the realm of computer vision. It is a powerful model that has shown promising results in image classification tasks.

The ViT model operates on an image by dividing it into a sequence of fixed-size patches. Each patch is then flattened and linearly projected into a vector representation. These patch embeddings, along with learnable position embeddings, are fed into a transformer model, which consists of multiple self-attention layers and feed-forward layers.

During training, the ViT model is typically pretrained on a large-scale dataset, such as ImageNet, using a self-supervised learning approach. This pretraining stage helps the model learn generic visual representations. After pretraining, the model can be fine-tuned on a smaller labeled dataset specific to the target task, such as image classification on a specific set of classes.

Fine-tuning involves updating the model's parameters using the labeled data from the target task, while retaining the knowledge learned during pretraining. This process allows the model to adapt its visual representations to the specific characteristics of the target dataset and improve its performance on the target task.

Fine-tuning a ViT model involves selecting an appropriate learning rate, optimizer, and training schedule. The process typically involves training the model on the target dataset for a certain number of epochs, adjusting the learning rate and other hyperparameters as necessary, and monitoring the model's performance on a validation set. It's also common to use techniques such as data augmentation and regularization to improve the model's generalization ability.

By fine-tuning a pretrained ViT model, you can leverage the knowledge learned from a large-scale dataset and achieve good performance even with limited labeled data. However, the success of fine-tuning depends on having a target dataset that is similar enough to the pretrained data and having enough labeled examples for the target task.

It's worth noting that the availability of pre-trained ViT models and specific fine-tuning techniques may vary based on the deep learning frameworks and libraries you are using.

## 5   Results

e conducted preliminary experiments on the dataset by using Machine Learning (ML), Deep Learning (DL), and Vision Transformer (ViT) Techniques. The consensus among raters regarding internal validation was measured by the exact Accuracy score.

### 5.1   Machine Learning

The machine learning models were trained using the structured 1D data. Multiple models were used for this process, and different results were obtained. The Gaussian classifier achieved the highest accuracy score, with 65% of the images correctly classified. The structured dataset used for this model was generated from RGB and grayscale images. That being said, both the RGB and grayscale structured datasets were responsible for achieving the highest accuracy score. We can assume that the reason for the Gaussian classifier being the best classifier is because the distributions of both the RGB and grayscale datasets are similar. In both cases, the distribution follows a Gaussian distribution, skewed in the positive direction (i.e., skewed to the right side). Keeping this in mind, the Gaussian classifier works best for data with a Gaussian distribution. Therefore, we can conclude that the reason for achieving the highest accuracy using the Gaussian classifier is the Gaussian distribution of the data as seen at Fig. 5 and Fig. 6.

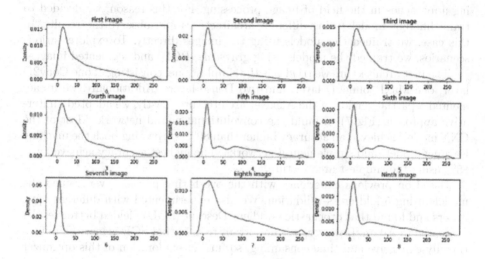

**Fig. 6.** Gaussian distribution of the first nine images of the Grayscale dataset

## 5.2   Deep Learning

We used the images directly for the deep learning models. In this scenario, compared with the machine learning approach, we are shifting from working with structured to working with unstructured data. We know for a fact that neural networks, especially convolutional neural networks, work well for image classification in most cases.

Before training a CNN, we wanted to reuse our structured dataset and see how a basic neural network would classify the data, and whether there would be a significant difference compared to the results from the machine learning models. For that reason, we built a simple neural network for this purpose and proceeded to fit the structured (1D) dataset to it. The first obstacle we faced was overfitting, which was expected due to having a large number of columns and a relatively small number of rows. We considered adding more Dropout layers, which randomly set input units are randomly set to 0 at each step with a frequency of "rate", aiming to mitigate overfitting. However, we were unsuccessful in reducing the overfit. Another approach we attempted was regularization, but it also did not effectively reduce the overfit. The experiments we conducted yielded different accuracy scores, but the best result came from the HOG dataset, with an accuracy score of 72%. Out of 50 epochs, the best fit was achieved by the 9th epoch, after which the model started to overfit. While this accuracy is not sufficiently high, it is higher than that of the models obtained through machine learning. Therefore, we can conclude that a neural network model yielded better results than a machine learning model for the same set of datasets.

The convolutional neural network (CNN) is known as one of the most promising approaches in the field of image processing. For this reason, we decided to train different models using different architectures and observe the results. In this case, we trained the models using the images directly. To explore various scenarios, we trained the models using grayscale, RGB, and segmented images.

The first architecture we tried was the simplest one, consisting of one Conv2D layer, one MaxPooling2D layer, and two Dense layers. Additionally, we implemented a partial version of the AlexNet architecture. Lastly, we adopted an iterative approach (Fig. 7) to build the convolutional neural network. Most of the CNN models achieved an accuracy higher than 80%, surpassing both the machine learning models and simple neural networks. The highest accuracy achieved was 85% using the simplest architecture.

Based on previous experience with the overfitting problem, we trained the models using 5-fold cross-validation. We also experimented with different optimizers and found that Stochastic Gradient Descent (SGD) yielded better results compared to Adam. However, it is important to note that SGD is more computationally expensive and time-consuming, so further exploration of this optimizer could be considered as future work.

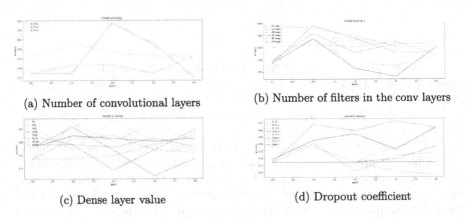

(a) Number of convolutional layers          (b) Number of filters in the conv layers

(c) Dense layer value          (d) Dropout coefficient

**Fig. 7.** Iteratively building of a CNN model architecture

## 5.3   ViT - Vision Transformer

The Transformer architecture has made a significant impact on the field of Natural Language Processing (NLP), inspiring researchers to develop more models based on Transformers. One such model is ViT, which stands for Vision

Transformer. ViT is a Transformer-based model designed specifically for image processing tasks. It is freely available and offered by the company HuggingFace.

To utilize this model, we needed to create a HuggingFace hub dataset to access and read the images. ViT was trained using our MRI images to enhance its understanding. However, we encountered another overfitting issue. ViT achieved an accuracy of 46%, which is considered poor and the worst performance thus far.

## 6    Conclusion and Future Work

In summary, the study underscores the significant potential of machine learning (ML) and deep learning (DL) models in predicting knee meniscus damage from MRI scans. The utilization of the MRNet dataset, in conjunction with different approaches including grayscale, RGB, and segmented images enriched with features derived from Histogram of Oriented Gradients (HOG) and Scale-Invariant Feature Transform (SIFT), has demonstrated encouraging results. Our ML and DL models have matched, and in some instances, exceeded the diagnostic proficiency of general medical doctors. These findings reiterate the capability of ML and DL models to provide swift preliminary results following MRI exams, augmenting the quality of MRI diagnoses. Particularly in environments lacking specialist radiologists, the integration of these models could significantly boost the reliability and accuracy of MRI interpretation for knee meniscus damage.

A significant body of work affirms that machine learning and deep learning methodologies are frequently correct ($¿9\%$) in image diagnosis in uncontrolled environments, and almost invariably ($¿95\%$) assist physicians in making decisions [4, 20, 24]. The findings of our investigations align with these studies.

This study opens several avenues for further research. Firstly, the investigation can be broadened to include other joint-related injuries and conditions, which could further validate the general applicability and robustness of the proposed ML and DL models. Secondly, the current study was limited by the use of the MRNet dataset. Expanding the dataset to include a more diverse population sample could improve the models' performance and generalizability. Additionally, integrating more sophisticated and diverse feature extraction methods might enhance the model's diagnostic capabilities. As the dataset size increases in real-life applications, one might need clusters on cloud resources customized for the data processing requirement at hand, while optimizing total cost of ownership [29].

Future studies should also consider exploring the possibility of real-time application of these models in clinical practice, addressing questions around implementation, acceptance, and ethical considerations. The integration of AI technologies into health care holds tremendous promise, but the transition must be managed carefully, with ongoing investigation to ensure patient safety, data security, and diagnostic accuracy. Finally, a cost-effectiveness analysis would be a valuable contribution, as the potential benefits of ML and DL models must outweigh the costs of their development, validation, and deployment.

**Acknowledgments.** This work is partilly funded by FCT/MEC through national funds and co-funded by FEDER—PT2020 partnership agreement under the project **UIDB/50008/2020**. This work is also funded by FCT/MEC through national funds and co-funded by FEDER—PT2020 partnership agreement under the project **UIDB/00308/2020**.

The work presented in this paper was partially financed by the University of Sts. Cyril and Methodius in Skopje, Macedonia, Faculty of Computer Science and Engineering.

# References

1. Corizzo, R., Dauphin, Y., Bellinger, C., Zdravevski, E., Japkowicz, N.: Explainable image analysis for decision support in medical healthcare. In: 2021 IEEE International Conference on Big Data (Big Data), pp. 4667–4674 (2021)
2. Maresova, P., et al.: Health-related ICT solutions of smart environments for elderly-systematic review. IEEE Access **8**, 54574–54600 (2020)
3. Ferreira, F., et al.: Experimental study on wound area measurement with mobile devices. Sensors **21**(17), 5762 (2021)
4. Fritz, B., Yi, P., Kijowski, R., Fritz, J.: Radiomics and deep learning for disease detection in musculoskeletal radiology: an overview of novel MRI- and CT-based approaches. Invest. Radiol. **58**(1), 3–13 (2023)
5. Hegde, A., George, R.M., Ranjith, H.: Detection and classification of knee osteoarthritis using texture descriptor algorithms. In: Intelligent Interactive Multimedia Systems for E-Healthcare Applications, pp. 151–166. Apple Academic Press (2022)
6. Senter, C., Hame, S.L.: Biomechanical analysis of tibial torque and knee flexion angle: implications for understanding knee injury. Sports Med. **36**, 635–641 (2006)
7. Lien-Iversen, T., Morgan, D.B., Jensen, C., Risberg, M.A., Engebretsen, L., Viberg, B.: Does surgery reduce knee osteoarthritis, meniscal injury and subsequent complications compared with non-surgery after ACL rupture with at least 10 years follow-up? A systematic review and meta-analysis. Br. J. Sports Med. **54**(10), 592–598 (2020)
8. Allum, R.: Complications of arthroscopic reconstruction of the anterior cruciate ligament. J. Bone Joint Surg. **85**(1), 12–16 (2003)
9. Renström, P.A.: Knee pain in tennis players. Clin. Sports Med. **14**(1), 163–175 (1995)
10. O'Brien, M.S., McDougall, J.J.: Age and frailty as risk factors for the development of osteoarthritis. Mech. Ageing Dev. **180**, 21–28 (2019)
11. Adams, B.G., Houston, M.N., Cameron, K.L.: The epidemiology of meniscus injury. Sports Med. Arthrosc. Rev. **29**(3), e24–e33 (2021)
12. Novriansyah, R., Kusuma, F.A.: Knee pain due to loose body in the knee joint: a case report in Dr. Kariadi general hospital Semarang. Med. Hospit.: J. Clin. Med. **9**(3), 378–382 (2022)
13. Sharma, L.: Osteoarthritis of the knee. N. Engl. J. Med. **384**(1), 51–59 (2021)
14. Paxton, E.S., Stock, M.V., Brophy, R.H.: Meniscal repair versus partial meniscectomy: a systematic review comparing reoperation rates and clinical outcomes. Arthrosc.: J. Arthrosc. Relat. Surg. **27**(9), 1275–1288 (2011)
15. Siouras, A., et al.: Knee injury detection using deep learning on MRI studies: a systematic review. Diagnostics **12**(2), 537 (2022)

16. Liu, F., et al.: Fully automated diagnosis of anterior cruciate ligament tears on knee MR images by using deep learning. Radiol.: Artif. Intell. **1**(3), 180091 (2019)

17. Sayegh, E.T., Matzkin, E.: Classifications in brief: the international society of arthroscopy, knee surgery, and orthopaedic sports medicine classification of meniscal tears. Clin. Orthop. Relat. Res.® **480**(1), 39–44 (2022)

18. Li, Z., et al.: Deep learning-based magnetic resonance imaging image features for diagnosis of anterior cruciate ligament injury. J. Healthc. Eng. **2021** (2021)

19. Petrovska, B., Zdravevski, E., Lameski, P., Corizzo, R., Štajduhar, I., Lerga, J.: Deep learning for feature extraction in remote sensing: a case-study of aerial scene classification. Sensors **20**(14), 3906 (2020)

20. Roblot, V., et al.: Artificial intelligence to diagnose meniscus tears on MRI. Diagn. Interv. Imaging **100**(4), 243–249 (2019)

21. Rizk, B., et al.: Meniscal lesion detection and characterization in adult knee MRI: a deep learning model approach with external validation. Phys. Med. **83**, 64–71 (2021)

22. Fritz, B., Fritz, J.: Artificial intelligence for MRI diagnosis of joints: a scoping review of the current state-of-the-art of deep learning-based approaches. Skeletal Radiol. **51**(2), 315–329 (2022)

23. Javed Awan, M., Mohd Rahim, M.S., Salim, N., Mohammed, M.A., Garcia-Zapirain, B., Abdulkareem, K.H.: Efficient detection of knee anterior cruciate ligament from magnetic resonance imaging using deep learning approach. Diagnostics **11**(1) (2021)

24. Bien, N., et al.: Deep-learning-assisted diagnosis for knee magnetic resonance imaging: development and retrospective validation of MRNet. PLOS Med. **15**(11), 1–19 (2018)

25. Ojala, T., Pietikainen, M., Maenpaa, T.: Multiresolution gray-scale and rotation invariant texture classification with local binary patterns. IEEE Trans. Pattern Anal. Mach. Intell. **24**(7), 971–987 (2002)

26. Pape, J.-M., Klukas, C.: 3-D histogram-based segmentation and leaf detection for rosette plants. In: Agapito, L., Bronstein, M.M., Rother, C. (eds.) ECCV 2014. LNCS, vol. 8928, pp. 61–74. Springer, Cham (2015). https://doi.org/10.1007/978-3-319-16220-1_5

27. Lowe, D.G.: Distinctive image features from scale-invariant keypoints. Int. J. Comput. Vision **60**(2), 91–110 (2004)

28. Jain, A.K., Farrokhnia, F.: Unsupervised texture segmentation using gabor filters. Pattern Recogn. **24**(12), 1167–1186 (1991)

29. Grzegorowski, M., Zdravevski, E., Janusz, A., Lameski, P., Apanowicz, C., Slezak, D.: Cost optimization for big data workloads based on dynamic scheduling and cluster-size tuning. Big Data Res. **25**, 100203 (2021)

# Image Classification Using Deep Neural Networks and Persistent Homology

Petar Sekuloski[✉] and Vesna Dimitrievska Ristovska

Faculty of Computer Science and Engineering, Ss. Cyril and Methodius University, Rugjer Boshkovikj, 16, 1000 Skopje, Macedonia

{petar.sekuloski,vesna.dimitrievska.ristovska}@finki.ukim.mk

**Abstract.** Persistent Homology (PH), a key tool in Topological Data Analysis (TDA), has gained significant traction in Machine Learning and Data Science applications in recent years. By combining techniques from algebraic topology, statistics, and computer science, PH captures the topological characteristics of datasets. This study aims to propose new classification models that integrate deep learning and Persistent Homology, exploring the impact of PH on model performance. Additionally, a transfer learning approach incorporating pre-trained networks and topological signatures is evaluated. Real-world datasets are used to assess the effectiveness of these models. The findings contribute to understanding the role of Persistent Homology in improving classification models, bridging the gap between deep learning, topological analysis, and practical data analysis. The performance of the models that include topological signatures showed better performance than the models that do not.

**Keywords:** Persistent Homology · Image Classification · Topological Data Analysis · Computational Topology · Deep Learning

## 1  Introduction

Image Classification is a complex task, especially when dealing with objects of the same category that exhibit variations in their appearance and pixel values. Traditional metrics that rely solely on measuring the distance between pixel values struggle to effectively capture the similarities between such images. This limitation has prompted researchers to explore alternative approaches, such as topology, to uncover hidden patterns that are invariant to common transformations.

Topology, as a branch of mathematics, provides a framework for studying the properties of spaces and their underlying structure. By focusing on the relationships and overall structure of an image, rather than pixel values alone, topology can reveal meaningful patterns that persist across different transformations like rotations, translations, and warping. This makes it a valuable tool in the context of Image Classification, where objects can appear in various orientations and positions.

One specific technique within applied topology that has gained attention in image analysis is Persistent Homology. It analyzes the presence of continuous regions and holes

M. Mihova and M. Jovanov (Eds.): ICT Innovations 2023, CCIS 1991, pp. 156–170, 2024.
https://doi.org/10.1007/978-3-031-54321-0_11

in an image and studies how these topological features evolve across different spatial scales. By examining the persistence of these features, researchers can extract robust image descriptors that capture important characteristics of the objects being classified. In this work, we propose deep learning models for image classification based on topological signatures and we evaluate that models on real world datasets.

The rest of the paper is organized as follows. Given the review of some research results in the field of TDA and ML especially in image classification in Sect. 2. Some mathematical concepts are discussed in Sect. 3. The methodology and proposed models are given in Sect. 4. The evaluation results are given in Sect. 5, together with a discussion on the models and the obtained results in Sect. 6. Finally, the conclusion of the research as well as the directions for future work are given in Sect. 7.

## 2 Related Work

In the research paper [1], authors explore the analysis of shape characteristics and categorization using a mathematical concept persistent homology. They apply this concept to two related constructions on geometric objects. By combining geometry and topology, this approach allows them to differentiate and classify shapes effectively. Using persistent homology, the authors in [1] extract a shape descriptor known as a barcode, which consists of a finite set of intervals. These intervals provide valuable information about the shape and its geometric properties. They define a metric that measures the similarity between different sets of intervals, allowing us to establish a continuous invariant for shapes.

To demonstrate the effectiveness of the methodology, they conduct thorough analyses on families of mathematically defined shapes, examining how different parameters impact their geometric properties.

The authors in [2] extend and modify the cubical complexes to create dual filtered cell complexes. By doing so, they establish a general connection between the persistent homology of these dual filtered cell complexes. They also investigate how making various modifications to a filtered complex affects the persistence diagram, which is a key representation of the persistent homology. Applying these findings to images, they develop a method to transform the persistence diagram obtained using one type of cubical complex into a persistence diagram for the other construction. This means that software used for computing persistent homology from images can be easily adapted to produce results for either of the two cubical complex constructions, without the need for additional low-level code implementation. In other words, this work simplifies the process of obtaining persistent homology results from images, allowing for greater flexibility and convenience in analyzing digital image data.

The paper [3] combines three different computational topology methods, namely effective homology, persistent homology, and discrete vector fields, to develop algorithms for homological digital image processing. These algorithms have been implemented as extensions of the Kenzo system, and their performance has been evaluated using real images obtained from a publicly available dataset. By integrating these computational topology techniques, the algorithms aim to analyze and process digital images based on their underlying topological features. The implemented algorithms have been tested on various digital images, and the results have demonstrated their effectiveness in processing and analyzing these images. This research contributes to the field of homological digital image processing by providing practical algorithms that leverage computational topology methods to extract meaningful topological and geometric information from digital images.

In the study [4], the authors propose an algorithm based on cubical homology specifically designed to extract topological features from 2D images. Cubical homology utilizes a collection of cubes to compute the homology, aligning well with the grid-like structure of digital images.

Although TDA is a relatively new field, in the last 4 years there have been a large number of scientific papers involving topological data analysis and it has been successfully applied in a number of fields. Some of the areas covered by the application of TDA are: gene expression, neural network analysis, chemoinformatics, time series prediction, cancer detection, cyber security, eco-informatics, natural language processing, sound processing, face recognition, analysis and time series prediction, stability of dynamical systems, image segmentation, sensor networks, complex networks, banking, noise detection, signal processing, bioinformatics, and many others [6–17].

## 3 Mathematical Background

In this section we will give a short introduction to the key mathematical concepts. More on these topics can be found in [16, 17].

Homology is a mathematical concept that establishes a connection between sequences of algebraic objects and topological spaces. The motivation behind introducing homology groups is to capture and quantify the presence of holes in a topological space. By analyzing the boundary operators and their action on chains, we can detect cycles (chains without boundaries) and boundaries (chains whose boundaries are cycles). The homology groups are then constructed by organizing these cycles and boundaries into equivalence classes. Each homology group corresponds to a specific dimension and captures a different aspect of the space's topology. The zeroth homology group counts the connected components of the space, while higher-dimensional homology groups reveal information about higher-dimensional holes or voids present in the space. The concept of homology groups provides powerful tools for distinguishing between shapes and spaces based on their topological properties. Two spaces are considered homotopy equivalent if their homology groups are isomorphic, meaning they share the same topological characteristics. By comparing the homology groups of different shapes, we can classify them into distinct topological classes. Furthermore, the computation of homology groups is achieved through various techniques such as simplicial homology, singular homology, or cellular homology. These methods involve constructing suitable chain complexes, applying boundary operators, and utilizing linear algebraic techniques to determine the homology groups.

The analysis in Persistent Homology involves constructing a filtration, which is a sequence of nested subsets of the dataset. As the filtration progresses, new topological features emerge, existing ones change, and some may disappear. The persistence of a feature is measured by the interval of filtration values during which it exists.

Persistent Homology is typically represented using persistence diagrams or barcodes, which provide a visual representation of the birth and death of topological features. These diagrams encode information about the lifespan and significance of different topological signatures.

**Definition 1.** A filtered cell complex $(X, F)$ is a cell complex $X$ together with a monotonic function $f : X \rightarrow \mathbb{R}$. A linear ordering $\sigma_0, \sigma_1, \sigma_2, \ldots, \sigma_n$ of the cells in $X$, such that $\sigma_i \preccurlyeq \sigma_j$ implies $i \leq j$, is compatible with the function $f$ when

$$f(\sigma_0) \leq f(\sigma_1) \leq f(\sigma_2) \leq \cdots \leq f(\sigma_n)$$

To represent the underlying structures of these images, CW-complexes are employed, which are generalizations of simplicial complexes. CW-complexes allow for the inclusion of cells that are not necessarily simplices, meaning they can take the form of various shapes such as cubes instead of tetrahedra. In this particular study, regular CW-complexes are utilized.

Since digital images can be represented as matrices or grid structures, it is more suitable to work with cell complexes rather than simplicial complexes. Cell complexes provide a more flexible framework for capturing the structures present in the images. By using cell complexes, the researchers are able to analyze and extract topological features from the images in a meaningful and computationally efficient manner.

By applying Persistent Homology to these cell complexes derived from the digital images, the study aims to track the evolution and persistence of topological characteristics across different scales or thresholds. This approach allows for a comprehensive analysis of the image structures, including the identification of significant regions, holes, or other topological features that may be relevant for image segmentation, classification, or other image analysis tasks.

Persistent homology is a mathematical framework that extends the concept of homology to analyze a filtration, which is a nested sequence of topological spaces. Homology groups, denoted as $H_k(X)$, provide information about the number of k-dimensional "holes" in a topological space $X$, where $k$ represents the dimension.

For example, $H_0(X)$, detects the connected components or "pieces" that form $X$, while $H_1(X)$, detects cycles or loops within $X$. The rank of the homology groups, known as the Betti numbers, represents the number of k-dimensional holes in $X$ and uniquely determines $H_k(X)$, for each $k$. In the context of analyzing cubical complexes, which is the case here, you can refer to reference [16] for detailed information on their homology.

## 4 Methodology

### 4.1 Constructing Cell Complexes from a Digital Image and Computing Persistent Homology

The first step to apply a topological method to a data set is to build a 'shape' (formally, a topological space) from the data points and some measure of similarity or distance between them. We build a topological space by gluing together basic building blocks, typically triangles or squares, and their higher-dimensional counterparts, with the data points as vertices. For images, pixels on a rectangular grid, using squares, cubes, etc., is the most natural option, resulting in what we call a cubical complex. More precisely, a cubical complex is a topological space made of any number of n-cubes, 0-cubes are points, 1-cubes are edges, 2-cubes are squares, 3-cubes are cubes, etc., for different $n$, glued together along their faces, the vertices of an edge, the sides of a square or a cube, etc., and embedded in $\mathbb{R}^m$, $n \leq m$. A cubical complex is a topological space, or, in our case, we construct a topological space from a digital image.

The process starts by selecting a threshold intensity level. Pixels in the image with intensities below this threshold are included in the complex, while those above the threshold are excluded. Adjacent pixels are then connected to form edges in the complex. In the case of a 2-dimensional image, squares are filled in by connecting the edges of adjacent pixels. For 3-dimensional images, cubes are filled in similarly.

By increasing the threshold intensity level, more pixels are included in the complex, leading to the formation of larger complexes. This process results in a sequence of increasingly larger complexes, which is referred to as a filtration, formally defined in previous section. The filtration captures the gradual inclusion of pixels based on their intensity levels. More on constructing cubical complexes can be found in [2].

The key phase of our methodology is construction of cubical complexes and computing persistent homology, see Fig. 1. Firstly, a cubic complex with a V-construction is constructed for each image in the dataset. We will mention that we will experiment with digital images consisted of grey-scale images. Then, homology is calculated on the complex that was obtained from the image. Once homology is computed, the Persistent Images can be determined. More information about Persistence Diagrams can be found in [18]. For the next step of this phase we will briefly introduce one more tool of Topological Data Analysis. That is Persistent Image. Persistent images are stable vector representations of persistent homology [15]. Persistent Images can be obtained from persistence diagrams and they are related to the persistent groups of a specific dimension. Since the images that were worked with in this study are two-dimensional, the dimension can only be either 0 or 1, so for every image we construct a Persistent image for dimension 0 and Persistent image for dimension 1, denoted with $PI_0$ and $PI_1$ in Fig. 1. This is actually the one of the main objectives of this work, to observe how the topological features extracted from the images will influence the classification process with the proposed network architectures in this work.

## 4.2 Network Architectures

In this section we will describe our proposed network architectures which specifications are shown on Fig. 2 and Fig. 3. First two of them are simple deep neural networks without topological signatures shown in Fig. 2. The second two, shown in Fig. 3, are proposed neural networks that use topological signatures based on Persistent Homology, each of them consist of three sub-parts which are deep networks, Part 1, Part 2 and Part 3, each with different data input. The preprocessed greyscale digital image (adapted for the input size) is denoted by $I$ and serves as the input for the Part 1. On the other hand, the data that we constructed for every original image that capture the topological signatures: the persistent image for dimension 0 ($PI_0$) as the input for the Part 2 and the

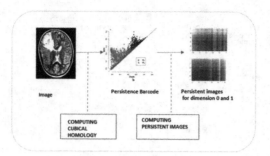

**Fig. 1.** Computing of Persistent Images

persistent image for dimension 1 ($PI_1$) as the input for the Part 3. Each output of the three sub-parts is concatenated in one vector. In the next step, for the experiments, we use additional activation layer for the vector that is obtained previously but this layer can be changed with any classifier. Also in our experiments we will use a model proposed in [9], where for the classification we have one part: classifier with the concatenated vector obtained from the original image and the persistent images as an input. We will denoted the model as $NA_0$. $NA_1$ and $NA_2$ are simple deep neural networks without Part 2 and Part 3. We used them for classification without topological signatures. In $NA_1$, we use a simple CNN with 4 convolutional layers and two dense layers. After convolutional layers they go through flatten layer and at the end we use softmax as the last layer. $NA_2$ is a simple dense neural network.

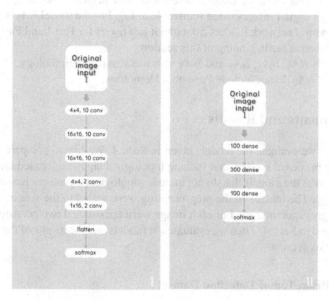

**Fig. 2.** Models without topological features. (I) $NA_1$ and (II) $NA_2$

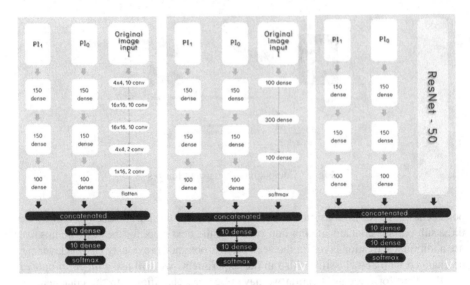

**Fig. 3.** Models with topological features. (III) $NA_3$, (IV) $NA_4$ and (V) $NA_5$

$NA_3$ is consisted of Part 1 - which is neural network $NA_1$ without two last layers dense and softmax and Part 2 and Part 3 as topological parts described above. $NA_4$ is consisted of Part 1 - which is neural network $NA_2$ without two last layers dense and softmax and Part 2 and Part 3 as topological parts described above.

As a part of this work we liked to investigate how persistent homology will affect classification of digital images using transfer learning. For the model $NA_5$ we used pre-trained deep neural network ResNet-50 without last layers for Part 1 and Part 2 and Part 3 as we described at the beginning of this section.

The models $NA_0$, $NA_3$, $NA_4$ and $NA_5$ also use Persistent Homology. On the other hand, $NA_1$ and $NA_2$ do not include Persistent Homology.

## 5   Experiments and Results

In this section we evaluate the models given in Sect. 4. We like to compare the performance of the proposed models that include topological signatures based on Persistent Homology versus the models that do not include topological signatures based on Persistent Homology. The images in the preprocessing were scaled to the size of $128 \times 128$ one channel greyscale images. For each image were constructed two persistent images, as we described in Sect. 4. Then we evaluate the models and the results of the evaluation are given in this section.

### 5.1   Br35::Brain Tumor Detection Dataset

Br35H::Brain Tumor Detection [19] is a public dataset containing 3000 patient images obtained by magnetic resonance imaging of the head region. It is intended for brain tumor detection. The images are divided into two classes: 1500 of the images belong to the "No"-0 class, the class indicating that the images are obtained from patients which do not have a tumor, and the "Yes"-1 class, which indicates that the images are obtained from patients who have a tumor.

For the $NA_0$ we get accuracy of 0.924. For more details of the evaluation of this model you can see [9].

The accuracy using $NA_1$ is 0.721. The values for other used metrics are given in Table 1.

**Table 1.**  Results of the evaluation of $NA_1$

| Class | Precision | Recall | f1-score |
|---|---|---|---|
| 0 | 0.7211 | 0.7125 | 0.6981 |
| 1 | 0.7159 | 0.6812 | 0.6745 |
| Average | **0.7185** | **0.6985** | **0.6863** |

The accuracy using $NA_2$ is 0.687. The values for other used metrics are given in Table 2.

**Table 2.**  Results of the evaluation of $NA_2$

| Class | Precision | Recall | f1-score |
|---|---|---|---|
| 0 | 0.7054 | 0.6254 | 0.5478 |
| 1 | 0.7521 | 0.6457 | 0.6354 |
| Average | **0.7185** | **0.6356** | **0.5916** |

The accuracy using $NA_3$ is 0.939. The values for other used metrics are given in Table 3.

**Table 3.**  Results of the evaluation of $NA_3$

| Class | Precision | Recall | f1-score |
|---|---|---|---|
| 0 | 0.9612 | 0.8901 | 0.93254 |
| 1 | 0.9103 | 0.9621 | 0.94561 |
| Average | **0.9358** | **0.9261** | **0.9391** |

The accuracy using $NA_4$ is 0.851. The values for other used metrics are given in Table 4.

The accuracy using $NA_5$ is 0.963. The values for other used metrics are given in Table 5.

**Table 4.** Results of the evaluation of $NA_4$

| Class | Precision | Recall | f1-score |
|---|---|---|---|
| 0 | 0.8455 | 0.8101 | 0.80145 |
| 1 | 0.8612 | 0.8751 | 0.7851 |
| Average | **0.8534** | **0.8426** | **0.7933** |

**Table 5.** Results of the evaluation of $NA_5$

| Class | Precision | Recall | f1-score |
|---|---|---|---|
| 0 | 0.9741 | 0.9454 | 0.9545 |
| 1 | 0.9612 | 0.9347 | 0.9781 |
| Average | **0.9676** | **0.9401** | **0.9663** |

The summary results of the evaluation of all used models are given in Table 6. The values for precision, recall and f1-scores are average values of these metrics for each of the classifiers we used. For $NA_0$ this values can be found in [9]. For the classifiers that use persistent homology there is PH on the right (for example $NA_0$ (PH)).

**Table 6.** Summary results of the evaluation of the models

| Classifier | Accuracy | Precision | Recall | f1-score |
|---|---|---|---|---|
| $NA_0$ (PH) | 0.924 | 0.9252 | 0.9035 | 0.9223 |
| $NA_1$ | 0.721 | 0.7185 | 0.6985 | 0.6863 |
| $NA_2$ | 0.687 | 0.7185 | 0.6356 | 0.5916 |
| $NA_3$ (PH) | 0.939 | 0.9358 | 0.9261 | 0.9391 |
| $NA_4$ (PH) | 0.851 | 0.8534 | 0.8426 | 0.7933 |
| $NA_5$ (PH) | 0.963 | 0.9676 | 0.9401 | 0.9663 |

## 5.2 Normal Peripheral Blood Dataset

The normal peripheral blood dataset [20] includes 17,092 images of separate cells. These images were captured using the CellaVision DM96 analyzer and are in the RGB color format. They have a size of 360 x 363 pixels. Trained clinical pathologists at the Hospital Clinic provided the labels for these images. The images are labeled in 8 labels, denoted from 0–7 for the purpose of the experiments. For the purpose of our experiment the images were transformed to grey-scale images and rescaled to the size of $128 \times 128$, before the used approach described in the Sect. 4. Also, this dataset was divided in to datasets: training set which contains 70% of the original dataset and testing set which contains 30% of the original dataset. This dataset is not balanced. The distribution of the samples into the classes in the training dataset and in the testing dataset is the same.

We used the same metrics for each of the models for the evaluation of the classification of the testing dataset.

For the $NA_0$ we get accuracy of 0.741. The values of other metrics are shown in Table 7.

**Table 7.** Results of the evaluation of $NA_0$

| Class | Precision | Recall | f1-score |
|-------|-----------|--------|----------|
| 0 | 0.7851 | 0.6877 | 0.6458 |
| 1 | 0.8757 | 0.5789 | 0.6787 |
| 2 | 0.6845 | 0.6454 | 0.5475 |
| 3 | 0.6788 | 0.6987 | 0.7236 |
| 4 | 0.7159 | 0.6812 | 0.6745 |
| 5 | 0.7231 | 0.7456 | 0.6342 |
| 6 | 0.6851 | 0.5145 | 0.6978 |
| 7 | 0.7212 | 0.4541 | 0.2541 |
| Average | **0.7337** | **0.6258** | **0.6070** |

The accuracy using $NA_1$ is 0.764. The values for other used metrics are given in Table 8.

**Table 8.** Results of the evaluation of $NA_1$

| Class | Precision | Recall | f1-score |
|-------|-----------|--------|----------|
| 0 | 0.5478 | 0.6124 | 0.6215 |
| 1 | 0.6241 | 0.5784 | 0.5478 |
| 2 | 0.5451 | 0.6872 | 0.7112 |
| 3 | 0.7342 | 0.6546 | 0.6731 |
| 4 | 0.6151 | 0.6215 | 0.6215 |
| 5 | 0.5428 | 0.5681 | 0.5998 |
| 6 | 0.6781 | 0.5478 | 0.5412 |
| 7 | 0.6872 | 0.6981 | 0.7421 |
| Average | **0.6218** | **0.6210** | **0.6323** |

The accuracy using $NA_2$ is 0.678. The values for other used metrics are given in Table 9.

**Table 9.** Results of the evaluation of $NA_2$

| Class | Precision | Recall | f1-score |
|-------|-----------|--------|----------|
| 0 | 0.6241 | 0.6752 | 0.6641 |
| 1 | 0.7215 | 0.7612 | 0.6734 |
| 2 | 0.6145 | 0.5568 | 0.5678 |
| 3 | 0.5426 | 0.6541 | 0.4521 |
| 4 | 0.5159 | 0.6212 | 0.6745 |
| 5 | 0.6428 | 0.6334 | 0.5987 |
| 6 | 0.5181 | 0.5147 | 0.5462 |
| 7 | 0.5887 | 0.6017 | 0.5549 |
| Average | **0.5960** | **0.6018** | **0.5915** |

The accuracy using $NA_3$ is 0.939. The values for other used metrics are given in Table 10.

**Table 10.** Results of the evaluation of $NA_3$

| Class | Precision | Recall | f1-score |
|-------|-----------|--------|----------|
| 0 | 0.9812 | 0.8724 | 0.8651 |
| 1 | 0.9125 | 0.9212 | 0.9451 |
| 2 | 0.9537 | 0.9001 | 0.8901 |
| 3 | 0.8619 | 0.8991 | 0.9287 |
| 4 | 0.9542 | 0.9768 | 0.9862 |
| 5 | 0.8912 | 0.8815 | 0.8762 |
| 6 | 0.9146 | 0.9247 | 0.9314 |
| 7 | 0.8964 | 0.8864 | 0.8697 |
| Average | **0.9207** | **0.9078** | **0.9116** |

The accuracy using $NA_4$ is 0.862. The values for other used metrics are given in Table 11.

The accuracy using $NA_5$ is 0.963. The values for other used metrics are given in Table 12.

The summary results of the evaluation of all used models are given in Table 13. The values for precision, recall and f1-scores are average values of these metrics for each of the classifiers we used. For the classifiers that use persistent homology there is PH on the right (for example $NA_0$ (PH)).

**Table 11.** Results of the evaluation of $NA_4$

| Class | Precision | Recall | f1-score |
|---|---|---|---|
| 0 | 0.8451 | 0.8545 | 0.8647 |
| 1 | 0.8489 | 0.8691 | 0.8357 |
| 2 | 0.8761 | 0.8712 | 0.8564 |
| 3 | 0.8564 | 0.5879 | 0.8796 |
| 4 | 0.8792 | 0.8792 | 0.8879 |
| 5 | 0.8567 | 0.8254 | 0.8351 |
| 6 | 0.8347 | 0.8221 | 0.8128 |
| 7 | 0.8451 | 0.8564 | 0.8116 |
| Average | **0.8541** | **0.8564** | **0.8116** |

**Table 12.** Results of the evaluation of $NA_5$

| Class | Precision | Recall | f1-score |
|---|---|---|---|
| 0 | 0.9987 | 0.9687 | 0.9567 |
| 1 | 0.9428 | 0.9645 | 0.9367 |
| 2 | 0.9412 | 0.9738 | 0.9271 |
| 3 | 0.9671 | 0.9125 | 0.8726 |
| 4 | 0.8876 | 0.9123 | 0.9461 |
| 5 | 0.9128 | 0.9537 | 0.8671 |
| 6 | 0.8256 | 0.8615 | 0.8762 |
| 7 | 0.8567 | 0.8739 | 0.8762 |
| Average | **0.9166** | **0.9276** | **0.9073** |

**Table 13.** Summary results of the evaluation of the models

| Classifier | Accuracy | Precision | Recall | f1-score |
|---|---|---|---|---|
| $NA_0$ (PH) | 0.741 | 0.7337 | 0.6258 | 0.6070 |
| $NA_1$ | 0.764 | 0.6218 | 0.6210 | 0.6323 |
| $NA_2$ | 0.678 | 0.5960 | 0.6018 | 0.5915 |
| $NA_3$ (PH) | 0.939 | 0.9207 | 0.9078 | 0.9116 |
| $NA_4$ (PH) | 0.862 | 0.8541 | 0.8564 | 0.8116 |
| $NA_5$ (PH) | 0.963 | 0.9116 | 0.9276 | 0.9073 |

# 6  Discussion

After analyzing the results, it can be concluded that the networks incorporating the Persistent Homology component achieved better performances compared to the networks without it. Specifically, when comparing the $NA_1$ model, a convolutional neural network, to the $NA_3$ model, which includes the same convolutional network along with topological signatures based on Persistent Homology, $NA_3$ showed significant improvement. It achieved at least a 30% improvement on the first dataset and at least a 22% improvement on the second dataset across all evaluation metrics.

The reason for the relatively worse performance of $NA_4$ compared to $NA_3$ can be attributed to the difference in architecture between the two models. In $NA_4$, Part 1 is implemented as a dense neural network, while in $NA_3$, Part 1 consists of a convolutional network.

Convolutional networks are widely regarded as more suitable for image classification tasks due to their ability to capture spatial relationships and local features in an image. Their convolutional layers apply filters across the input image, extracting relevant features at different scales. This characteristic allows convolutional networks to effectively learn hierarchical representations from image data.

On the other hand, dense neural networks, also known as fully connected networks, lack the spatial awareness and parameter sharing present in convolutional networks. Dense networks typically have all neurons connected to every neuron in the previous and subsequent layers, which can lead to a large number of parameters and difficulties in capturing spatial information effectively.

Also, it is important to note that the $NA_0$ model proposed in our previous work, based on Persistent Homology, performed worse than the other models based on topological signatures. This suggests that the models leveraging topological information outperformed the $NA_0$ model, which concatenated the flattened original image with the persistent images before inputting them in the neural network. When comparing the metrics of the model that includes topological signatures but flattens and concatenates the original image with the persistent images as a single input in a one-part neural network, it also showed worse performance than the other models utilizing persistent homology. This indicates that each stream of information, providing distinct features, enables more effective feature representation.

Additionally, as part of this study, we evaluated the performance of a transfer learning model that combines a pre-trained convolutional network, specifically ResNet-50, with the inclusion of Persistent Homology. This model demonstrated the best performance among all the models evaluated. This can be attributed to the fact that pre-trained models like ResNet-50 have consistently shown state-of-the-art performance in image classification tasks. The fusion of the pre-trained network with Persistent Homology resulted in high values for the evaluation metrics used in both datasets. These findings further emphasize the effectiveness of leveraging both pre-trained models and topological information to achieve superior performance in image classification tasks.

# 7   Conclusion and Further Work

The comprehensive analysis of the results leads to the compelling conclusion that the inclusion of Persistent Homology and topological signatures in the networks has proven to be highly advantageous for image classification tasks. The models that incorporated these components consistently demonstrated substantial improvements in performance across a range of evaluation metrics, surpassing the models that did not utilize them. Of particular significance is the remarkable performance achieved by the models that combined pre-trained convolutional networks with Persistent Homology. This fusion of pre-existing models with topological information showcased superior results, emphasizing the synergistic benefits derived from leveraging both sources of knowledge. By effectively capturing and integrating topological features, these models were able to enhance the representation of key characteristics and significantly improve the accuracy of image classification. These findings reaffirm the significance of topological analysis in augmenting the overall performance and effectiveness of the networks, underscoring its potential as a valuable tool in the realm of image classification.

Looking ahead, there are several avenues for further exploration in harnessing topological features for improved classification. One potential direction is to investigate the impact of topological features in more complex classifiers or develop intricate models that integrate them. By incorporating topological characteristics into advanced classification algorithms, a deeper understanding of their influence and potential for enhancing performance can be gained.

Additionally, during the training process, it may be beneficial to define a loss function that incorporates topological characteristics. This would enable the network to optimize its parameters with respect to these features, further emphasizing their importance and potentially leading to improved classification performance.

Moreover, there is an opportunity to delve into the study and development of input parameters for selected algorithms within the field of Topological Data Analysis (TDA). Exploring the influence of parameter choices on the obtained results for specific datasets can provide valuable insights into the optimal parameter settings and their impact on classification outcomes.

**Acknowledgement.** The research presented in this paper is partly supported by the Faculty of Computer Science and Engineering, at the Ss. Cyril and Methodius University in Skopje.

# References

1. Carlsson, G., Zomorodian, A., Collins, A., Guibas, L.: Persistence barcodes for shapes. Int. J. Shape Model. **2**(5), 99–110 (2004)
2. Bleile, B., Garin, A., Hesis, T., Maggs, K., Robins, V.: The persistent homology of dual digital image constructions. arXiv (2021). arXiv:2102.11397
3. Romero, A., Rubio, J., Sergeraert, F.: Effective persistent homology of digital images (2014) arXiv:1412.6154
4. Choe, S., Ramanna, S.: Cubical Homology-Based Machine Learning: An Application in Image Classification. Axioms (2022)

5. Pun, C.S., Lee, S.X., Xia, K.: Artif. Intell. Rev. **55**, 5169–5213 (2022)
6. Sekuloski, P., Dimitrievska Ristovska, V.: Application of persistent homology on bio-medical dataset - a case study. In: Mathematical Modeling - Proceedings, III International Scientific Conference, Bulgaria (2019)
7. Dimitrievska Ristovska, V., Sekuloski, P.: Mapper algorithm and its applications. In: Mathematical Modeling - Proceedings, III International Scientific Conference, Bulgaria (2019)
8. Sekuloski, P., Dimitrievska Ristovska, V.: Classification of digital images using topological signatures - a case study. In: Mathematical Modeling - Proceedings, III International Scientific Conference, Bulgaria (2022)
9. Sekuloski, P., Dimitrievska Ristovska, V.: A novel model for image classification based on Persistent Homology. Int. J. Sci. Res. (IJSR) (2022)
10. De Silva, V., Ghrist, R.: Coordinate-free coverage in sensor networks with controlled boundaries via homology. Int. J. Robot. Res. **25**, 1205–1222 (2006)
11. Don, A.P.H., Peters, J.F., Ramanna, S., Tozzi, A.: A topological view of flows inside the BOLD spontaneous activity of the human brain. Front. Comput. Neurosci. **14**, 34 (2020)
12. Carrière, M., Chazal, F., Ike, Y., Lacombe, T., Royer, M., Umeda, Y.: Perslay: a neural network layer for persistence diagrams and new graph topological signatures. In: Proceedings of the International Conference on Artificial Intelligence and Statistics (PMLR), Online, 26–28 August 2020, pp. 2786–2796 (2020)
13. Chung, M.K., Lee, H., DiChristofano, A., Ombao, H., Solo, V.: Exact topological inference of the resting-state brain networks in twins. Netw. Neurosci. **3**, 674–694 (2019)
14. Nicolau, M., Levine, A.J., Carlsson, G.: Topology-based data analysis identifies a subgroup of breast cancers with a unique mutational profile and excellent survival (2011)
15. Adams, H., et al.: Persistence Images: a stable vector representation of persistent homology. Found. Comput. Math. **18**, 1–35 (2018)
16. Hatcher, A.: Algebraic Topology. Cambridge University Press (2002)
17. Munkers, J.R.: Topology, vol. 2. Prentice Hall, Upper Saddle River (2000)
18. Zomorodian, A., Carlsson, G.: Computing persistent homology. Discret. Comput. Geom. **33**, 249–274 (2005)
19. Ahmedhamada: Brain Tumor Detection Dataset. https://www.kaggle.com/datasets/ahmedhamada0/brain-tumor-detection. Accessed 15 Aug 2022. Accessed 27 July 2022
20. Acevedo, A., Alférez, S., Merino, A., Puigví, L., Rodellar, J.: Recognition of peripheral blood cell images using convolutional neural networks. Comput. Methods Programs Biomed. **180**, 105020 (2019)

# Network Science

# Understanding Worldwide Natural Gas Trade Flow for 2017 to 2022: A Network-Based Approach

Jovana Marojevikj[1]($\boxtimes$), Ana Todorovska[1], Irena Vodenska[1,3], Lou Chitkushev[2], and Dimitar Trajanov[1,2]

[1] Faculty of Computer Science and Engineering, Ss Cyril and Methodiuos University, Skopje, North Macedonia
`jovana.marojevikj@students.finki.ukim.mk`,
`{ana.todorovska,dimitar.trajanov}@finki.ukim.mk`
[2] Computer Science Department, Metropolitan College, Boston University, Boston, USA
`lou@bu.edu`
[3] Administrative Sciences Department, Metropolitan College, Boston University, Boston, USA
`vodenska@bu.edu`

**Abstract.** Natural gas is a critical commodity in the global economy, and its trade dynamics and price movements are of significant interest, particularly during times of major economic disruptions. In this research, we investigate the natural gas trade from 2017–2022 using UN Comtrade data. Our goal is to identify patterns in countries' reliance on specific gas exporters and their strategies for reducing the risk of supply disruptions. To achieve this, we analyze trade flows between countries using graph theory methods and construct networks that illustrate these flows. Our findings indicate that the gas trade network has become more interconnected over time, suggesting increasing globalization. In addition, we create year-over-year (YoY) networks that capture changes in natural gas prices for each year. Our analysis shows that, while natural gas prices have generally increased over time, there was a decrease in the price of gas exports from the Russian Federation to Serbia and Armenia in 2022 compared to 2021. Our research provides insights into the evolution of the natural gas trade and its price fluctuations, and the proposed methodology can be extended to other globally important commodities.

**Keywords:** centrality measures · YoY graphs · network theory · natural gas · the UN Comtrade

## 1 Introduction

Natural gas serves as a vital resource, extensively utilized by numerous countries to meet their electricity and heating demands, making it a crucial commodity on the global stage. The trading of natural gas and oil is closely intertwined

Supported by Faculty of Computer Science and Engineering, Skopje, N. Macedonia.

M. Mihova and M. Jovanov (Eds.): ICT Innovations 2023, CCIS 1991, pp. 173–190, 2024.
https://doi.org/10.1007/978-3-031-54321-0_12

with the interconnections between economics, politics, resource allocation, and global events [5]. These factors play a crucial role in shaping the dynamics of the natural gas and oil markets and have significant implications for the global economy and international relations [18] [21]. Therefore, it becomes imperative to comprehensively investigate and understand the underlying networks formed by these gas trade transactions.

In this research, we present an extensive analysis of natural gas imports on a global scale, spanning the years 2017 to 2022, leveraging data from the United Nations Comtrade database [19] - a comprehensive platform providing comprehensive global annual trade statistics by product and trading partner. To better comprehend the complex flows of the natural gas trade, we employ a multi-faceted approach. Firstly, we visualize the networks on a year-by-year basis, enabling the observation of changes in interconnectivity and the identification of key players in the natural gas trade, including the largest importers and exporters. Subsequently, we employ graph theory methods, specifically centrality measures [11], which refer to a set of quantitative metrics used to assess the relative importance or prominence of individual nodes within a network, which captures the significance of a node or, in this case, a country, based on its structural position and influence within the graph. By evaluating a country's structural position and influence, we aim to identify the most prominent countries involved in natural gas exports, recognize nations overly reliant on a few exporters leading to bottlenecks and vulnerability in their imports, and determine countries with diversified sources of natural gas imports. Such analysis provides valuable insights and diversification strategies for nations.

Furthermore, we employ year-over-year (YoY) networks to analyze annual variations in prices by comparing data between consecutive years. By investigating fluctuations in prices, we aim to uncover underlying factors such as inflation, economic shocks, and political relations between partnering countries. These elements enhance the comprehension of price dynamics within the natural gas trade, aiding in recognizing potential risks and opportunities for the participating nations.

The primary goal of this research is to gain comprehensive insights into the dynamic evolution of the gas trade network over time and draw meaningful conclusions regarding its future trajectory. By shedding light on the interdependencies, vulnerabilities, and potential strategies for diversification in natural gas trading, our study aims to inform policy decisions and contribute to the sustainable development and stability of energy markets worldwide.

## 2   Literature Review

The global natural gas trade has been growing rapidly in recent years, with new trade routes emerging and traditional trade patterns shifting caused by periods of economic shocks worldwide, which has made natural gas trade dynamics an attractive subject to researchers.

Filimonova et al. (2022) [3] focused on potential network link predictions within the LNG market. Through quantitative analysis of the global LNG trade, the researchers found that the current network has a low degree of centrality and density, suggesting opportunities for expansion in LNG trade connections. The findings indicated that European countries are prominently positioned as a promising market for LNG. Geng et al. (2014) [4] conducted a dynamic analysis of the global natural gas trade network using a complex network theory approach. In their methodology, they propose examining the topological properties of the international gas trade network, such as out-degree, in-degree centrality, and clustering coefficient. They proceed with creating a minimum spanning tree model. They come to the conclusion that the world has entered an age in which many countries will increase their use of natural gas, demand will rise, and the increase in natural gas trade will strengthen among the three markets (North America, Europe, and Asia).

Halser and Paraschiv (2022) [6] focused on the German case and explored pathways to overcoming natural gas dependency on Russia. The authors analyzed the current energy mix of Germany and discussed alternative sources of natural gas supply. Additionally, they suggest policy measures such as short-term supply substitution and short-term demand reductions to diversify natural gas supply sources and reduce reliance on the Russian Federation.

To provide a comprehensive global perspective on the utilization of natural gas within the economy, Kan et al. (2019) [17] employed the multi-regional input-output (MRIO) approach. This method allowed them to trace the journey of natural gas consumption, starting from primary suppliers and extending to ultimate consumers. The outcomes of their analysis revealed a significant repositioning of gas flows dedicated to trade activities. The study's findings underscored that a substantial portion, amounting to 83% of the traded natural gas, is linked with intermediate trading stages. Notably, this flow originates predominantly from major primary suppliers such as Russia, the USA, and Western Asia. These suppliers funnel the natural gas to key final consumers, with prominent examples being the EU27 and East Asia.

When analyzing natural gas trade for countries under the Belt and Road Initiative (BRI), in the studies [20] and [9], the authors investigate through the construction of networks that incorporate pipeline natural gas (PNG) and liquefied natural gas (LNG). Complex network theory is employed to analyze the evolutionary patterns of the natural gas trade network. The findings of their research indicate that LNG trade along the BRI exhibits rapid growth, whereas the development of PNG trade remains relatively stable.

To assess and deliberate upon the consequences of worldwide liquefied natural gas (LNG) trade within a low-carbon framework, Lin and Brooks (2021) [12] utilized a partial equilibrium model. Consequently, the scholars examine how a structural economic model plays a crucial part in empirically analyzing the implications of a transition, such as a shift towards alternative energy sources.

In the context of post-COVID-19 conditions and the globalization of oil and natural gas trade, challenges and opportunities have emerged. Norouzi (2021) [13] discussed the implications of the pandemic on the global oil and gas markets, including disruptions in supply chains, reduced demand, and changes in trade patterns. The author highlights the importance of regulatory frameworks and strategies to ensure the resilience and stability of the global energy market. Suggesting a bifurcated approach, Jana and Ghosh (2022) [8] put forth a granular two-stage framework aimed at predicting natural gas futures behavior prior to and during the COVID-19 phases. The outcomes of their study highlight the remarkable efficacy of the proposed framework in deciphering patterns even when confronted with a restricted dataset. Lu et al. (2020) [7] provided a decade review of data-driven models for energy price prediction and discussed the advantages and limitations of various data-driven models. The study also discusses the potential applications of energy price prediction models in natural gas trading.

Overall, the literature suggests that natural gas dependency on Russia remains a significant challenge for many countries, but policy measures and diversification of supply sources can help to mitigate this risk. The transformation of LNG markets and the emergence of new trade routes also offer opportunities for natural gas trading. Additionally, dynamic analysis of the global natural gas trade network can help to identify the key drivers of natural gas trade flows, and data-driven models can provide insights into energy price prediction.

## 3   Data

The entirety of the data utilized in this research has been sourced from reputable and authenticated sources. The primary source is the UN Comtrade database, which represents the United Nations Trade Statistics database. More than 170 reporter countries/areas supply their annual trade statistics data categorized by commodities/service types and partner countries to the United Nations Statistics Division (UNSD)[1]. As a result, the UN Comtrade database stands as the largest collection of international trade data. In collaboration with organizations such as the International Trade Center[2], United Nations Statistical Division (UNSD), and the World Trade Organization (WTO), the World Bank[3] created the World Integrated Trade Solution (WITS)[4] software. The program empowers users to conveniently retrieve trade data and tariff information.

### 3.1   Data Retrieval

The Advanced Query's UN Comtrade data feature furnishes trade value and quantity data from UNSD. It facilitates the creation of intricate queries, encompassing numerous reporters, partners, products, trade flows, and years, all within a single query. For the purpose of this research, the advanced query constructed and executed is "Gas Trade Network: Import of natural gas in the gaseous state by country" with the following parameters:

---

[1] https://unstats.un.org/UNSDWebsite/about/.

[2] https://intracen.org/about-us.

[3] https://www.worldbank.org/en/home.

[4] https://www.wto.org.

- Reporters: Reporters are countries that import from the selected exporters. All of the reporters available in the Comtrade database were selected, a total of 120 unique countries.
- Products: The foundation of all WITS queries rests upon the nomenclature (or classification) principle, making it essential while utilizing WITS. The nomenclatures employed in WITS are globally accepted standard classifications that serve the objectives of trade, tariff, industry, and national income accounts. According to this Harmonized System[5], the natural gas commodity code is: 2711 21 - Petroleum gasses and other gaseous hydrocarbons, in a gaseous state, natural gas.
- Partners: Partners are countries that export the selected goods to the reporters. All of the partners available were selected, a total of 100 unique countries.
- Years: 2017 to 2022
- Trade flow: In a perfect world, country A reported imports from country B would match with country B's reported exports to country A. Consequently, this would make mirroring (using information from the partner when a country does not report its trade) a transparent and error-free process. According to Comtrade[6], for a given country, imports are usually recorded with higher accuracy due to various reasons listed by the UN statistics division[7], such as time lag between imports and exports, goods going via third countries, different trade systems between countries etc. In order to avoid mirroring of the data, the trade flow chosen in the advanced query for data retrieval was import flow.

## 3.2  Data Transformation and Clean Up

Initially, the data consisted of 1094 entities, with 111 unique reporter countries and 94 unique partner countries. However, like any other data sourced from various origins, it contained null and empty values, indicating the need for additional cleanup. Rows with null values for both quantity and trade value, as well as those involving quantities of natural gas, traded less than 1000kg, were removed from the data set as they lacked significance. Following the cleanup process, the data set comprised 900 entries, with 92 unique reporter countries and 79 unique partner countries.

## 4  Visual Data Analysis

The data is structured as a network comprising a "from" node, a "to" node, and an optional weight parameter representing the trade value associated with each edge connecting the nodes. Given the data's nature, Jaal[8], a Python-based

---

[5]  https://www.wcoomd.org/en/topics/nomenclature/overview.aspx.

[6]  https://wits.worldbank.org/wits/wits/witshelp/content/data_retrieval/T/Intro/ B2.Imports_Exports_and_Mirror.htm.

[7]  https://unstats.un.org/wiki/display/comtrade/Exports+of+a+country+not+coinc ide+with+imports+of+its+partner.

[8]  http://mohitmayank.com/jaal/.

**Fig. 1.** Gas trade network 2017          **Fig. 2.** Gas trade network 2022

interactive network visualization tool, is utilized. Jaal offers a range of features, including the ability to create directed graphs, weigh nodes and edges, and color them.

### 4.1   Global Natural Gas Trade and Its Evolution over the Years

When analyzing trade data, it is essential to obtain a comprehensive overview of the big picture, including the flows, nodes, and edges in a "zoomed-out" view. The primary focus is to examine how the network evolves over time.

The analysis of the gas trade network's connectivity proves to be an intriguing area of observation. Notably, the central region, representing Europe, exhibits greater interconnectivity compared to the left and right sides of the network. This observation suggests Europe's significant reliance on natural gas and highlights the influence of proximity and transportation channels in fostering interrelations within the region.

Another aspect of interest lies in the examination of connected components within the network. In graph theory, a connected component [16] refers to a sub-graph that is not part of any larger connected sub-graph. In Fig. 1, four distinct connected components can be identified. The largest component, situated on the rightmost side, encompasses the European region and its associated partners. The second component comprises Bolivia and Brazil, both located in South America, which explains their exclusion from the European connected component due to the complexities associated with natural gas transportation. Similarly, the third connected component consists of Pakistan and the United Arab Emirates. Lastly, a smaller connected component includes countries from the Sub-Saharan African region and the South American continent, namely Qatar, Chile, Argentina, and Uruguay.

When evaluating the evolution of the gas trade network from 2017 to 2022, noticeable changes become apparent. The 2022 network in Fig. 2, exhibits an increase in the number of nodes and edges, albeit with fewer and smaller connected components compared to the 2017 network. These findings indicate a more globalized gas trade with enhanced interconnectivity among partnering countries. Consequently, it can be inferred that countries have the opportunity to diversify their natural gas suppliers over time due to the increasing globalization and interdependence within the trade flow.

## 4.2    Most Significant Natural Gas Trade Flows

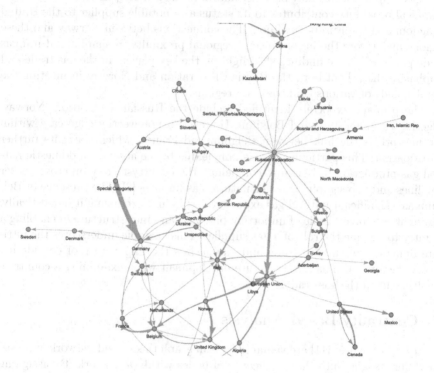

**Fig. 3.** Top 20% of the natural gas trade flows by country 2017–2022

In order to get further insight into the most involved countries and the flows of trade for the past five years, all of the data from the last six years was summed up. Due to the complexity, the size of the resulting network, and for visualization purposes, it was best to sort the data by trade value and separate only the top 20% of the trade flows.

The largest link that is initially noticeable in Fig. 3 is the one connecting "Special categories" to Germany. The partner code "Special Categories" (839), according to Comtrade, is used when a reporting country prefers not to disclose the partner breakdown[9]. This link carries the highest value in the entire graph and thus cannot be ignored. Another notable feature is that Germany, has six

---

[9] https://unstats.un.org/wiki/display/comtrade.

outward nodes, indicating that it has been exporting gas to partner countries in the past five years.

The subsequent notable link encompasses the connections between the Russian Federation and the European Union, as well as Italy. These particular links exhibit the highest weight within the network. The Russian Federation emerges as a dominant player in the gas trade, exporting gas to a total of 19 countries, as evidenced by the presence of 19 outward nodes. Norway assumes a significant role as a gas supplier, which may be lesser known to the general public. Norway's geographical proximity contributes to its status as a notable supplier to the United Kingdom and the European Union. The connections between Norway and these regions underscore the importance of regional proximity in shaping natural gas trade patterns. These findings shed light on the key players in the gas trade and emphasize the role of both the Russian Federation and Norway in meeting the gas demands of various countries and regions.

The primary exporters identified include the Russian Federation, Norway, Algeria, Turkmenistan, and Belgium. However, a noteworthy edge case within the network is the link between Belgium and France, which warrants further investigation. This particular case is unique due to the absence of domestic natural gas production in Belgium, compelling the country to rely on cross-border pipelines and sub-sea pipelines for natural gas imports [2]. The presence of Belgium as a significant natural gas exporter, despite not producing it domestically, highlights the pivotal role of importing pipelines and infrastructure in enabling a country to assume the role of a gas supplier. This example underscores the intricate dynamics of the gas trade and emphasizes the importance of considering diverse factors, such as infrastructure development when examining a country's involvement in the gas trade.

## 5   Centrality Based Analysis

Centrality measures [11] are valuable tools in graph theory and network analysis, providing insights into the significance of nodes within a network. By assigning numerical values or rankings to nodes based on their network positions, centrality measures facilitate the identification of important nodes in the graph [15]. In this research, we employ centrality measures to uncover key aspects of global natural gas trading networks, including dominant nodes, diversification strategies, and high dependence on specific suppliers. By focusing on the years 2017–2022, we calculate centrality measures for a comprehensive data set and present the top 10 scoring countries in the results. All measures for the centrality represented in the tables have been normalized using min-max normalization, meaning that the centrality of the most central node is set to 1, and the lowest to 0.

### 5.1   In Degree, Out Degree and Eigenvector Centrality

Degree centrality [10] is a measure of a node's importance based on the number of edges it has. As the degree increases, the graph's significance amplifies accordingly. Consequently, the in-degree corresponds to the total number of incoming connections directed towards the node, while the out-degree represents the total number of connections that the node establishes with other entities. In the case of trade, the highest out-degree centrality will have the country that is the biggest exporter of natural gas, while the country with the highest in-degree would be one that imports from the most different sources. The top ten nodes for in-degree and out-degree centrality are shown in Table 1.

Eigenvector centrality [1] gauges the influence of a node within a network by assigning relative scores to all nodes. It operates on the principle that connections to nodes with higher scores contribute more to the score of the focal node compared to connections to nodes with lower scores. Simply put, eigenvector centrality determines a node's significance based on the significance of its neighboring nodes. If a node is connected to or surrounded by highly significant nodes, it will have a high eigenvector centrality score, indicating its importance in the network. This centrality measure identifies nodes that have an impact on the entire network, not just those directly linked to them. A high score for a node automatically implies that its neighboring connected nodes also have high scores. The networkX implementation of eigenvector centrality takes into consideration the weights of the links when calculating the scores. Similarly to the degree centrality, this centrality measure goes deeper by paying regard to how well connected a neighbor node is by calculating how many links their connection has. This approach will reveal which nodes have the broadest reach in the network. The top ten nodes for eigenvector centrality are shown in Table 1.

**Table 1.** Top 10 nodes for in-degree, out-degree, and eigenvector centrality

| In-degree Centrality | | Out-degree Centrality | | Eigenvector Centrality | |
|---|---|---|---|---|---|
| Country | Score | Country | Score | Country | Score |
| European Union | 1 | United States | 1 | United States | 1 |
| United Arab Emirates | .889 | United Kingdom | .566 | Mexico | .874 |
| Bulgaria | .556 | Germany | .451 | Canada | .513 |
| Ukraine | .444 | Russian Federation | .422 | European Union | .011 |
| Switzerland | .333 | China | .393 | Bulgaria | .01 |
| Czech Republic | .222 | Italy | .191 | Czech Republic | .007 |
| Spain | .222 | The Netherlands | .162 | Morocco | .006 |
| Greece | .222 | France | .046 | Ukraine | .001 |
| Hungary | .111 | Austria | 0 | Brazil | 0 |
| Thailand | 0 | United Arab Emirates | .03 | Estonia | 0 |

## 5.2 Node Strength in and Node Strength Out

This metric represents the strength of the relations that the node has with its neighbors [14]. Given that the graph is directed, it would be necessary to calculate both the in-node strength, which will sum the weights of the links going into a node (import), and out-node strength, which will sum the weights of the link going out of a node (export). The higher the out-node strength means the node is a major exporter of natural gas, while a high value for in-node strength means that a country relies heavily on natural gas and imports the most of it. This information gives an insight into the nodes that control the information flow within the network of gas trade. The top ten nodes' in-node and out-node strength are shown in Table 2.

**Table 2.** Top 10 nodes for in-node and out-node strength

| In-node strength | | Out-node strength | |
|---|---|---|---|
| Country | Score | Country | Score |
| European Union | 1 | Russian Federation | 1 |
| Germany | .895 | Special Categories | .836 |
| Italy | .483 | Norway | .716 |
| Belgium | .362 | Algeria | .310 |
| United Kingdom | .298 | Belgium | 0 |
| China | .260 | Canada | 0 |
| France | .248 | Turkmenistan | 0 |
| United States | .154 | Azerbaijan | 0 |
| Mexico | 0 | United Kingdom | 0 |
| Hungary | 0 | United States | 0 |

## 6    Year-over-Year (YoY) Analysis

The primary objective of this section is to analyze the year-over-year changes in natural gas prices. To accomplish this, we calculate the difference in price per kilogram of gas between each year and the preceding year. This approach allows us to determine whether country A purchased natural gas from country B at a relatively cheaper or more expensive price in a given year (X) compared to the previous year (Y). Consequently, we generate five YoY networks, each representing the difference in prices between countries for the years 2018-2017, 2019-2018, 2020-2019, 2021-2020, and 2022-2021. By constructing these YoY networks, we aim to uncover patterns and trends in gas prices, enabling us to assess the dynamics of price fluctuations over time. This approach offers valuable insights into the evolving relationships between countries in terms of gas trade and the economic implications of price differentials. Through the examination of these networks, we can gain a comprehensive understanding of the changing landscape of natural gas prices, contributing to a deeper understanding of the factors influencing global gas trade dynamics.

### 6.1    2018 - 2017 Network

The initial graph under analysis pertains to the YoY difference between the years 2018 and 2017. The data frame from 2017 consisted of 82 nodes and 149

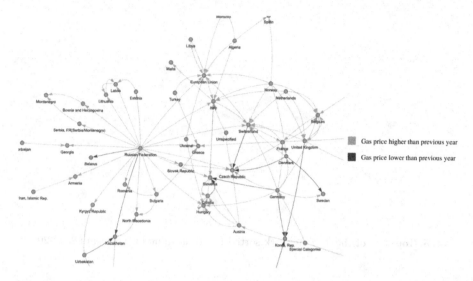

**Fig. 4.** Top 30% of the YoY network sorted by trade value for the years 2018-2017

edges, while the 2018 data frame comprised 91 nodes and 176 edges. Following the merging process, the resulting network consisted of 81 nodes and 132 edges. These findings indicate a high level of consistency, as most partner and reporter countries remained the same between 2017 and 2018. In Fig. 4, a notable observation is the predominance of trade flows with higher prices per kilogram of natural gas in 2018 compared to 2017. The total number of links with increased prices is 114. However, a few exceptions are identifiable - a total of 18 links, experiencing a decrease in price. Notably, these exceptions involve trade flows between neighboring countries or countries in close proximity. Specifically, the edges from the Russian Federation to Belarus and Kazakhstan, as well as from Croatia to Slovenia, Denmark, and Sweden, demonstrate this price decline pattern. The shared characteristic among these edges is their geographical proximity, which likely contributes to the observed price decrease phenomenon.

## 6.2    2019 - 2018 Network

The subsequent network under examination corresponds to the YoY network between the years 2019 and 2018. Prior to merging the two data frames, each data frame possessed distinct properties. The data frame for 2018 comprised 91 nodes and 176 edges, while the data frame for 2019 consisted of 91 nodes and 179 edges. Following the merging process, the resulting network revealed 86 nodes and 144 edges. This outcome further supports the notion that links between nodes tend to exhibit repeatability over the years, indicating predictability in the trade flows. In this network, 68% of the edges are colored orange, indicating a decrease in gas prices, while 32% of the edges are purple, representing the opposite scenario. This observation suggests that gas prices do not consistently

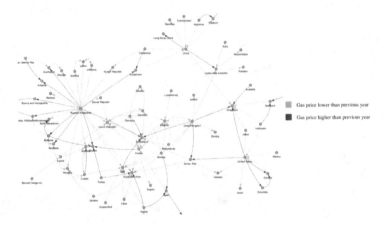

**Fig. 5.** Top 30% of the YoY network sorted by trade value for the years 2019-2018

follow a global trend but rather depend on the specific partnering countries and their relationships. These findings emphasize the significant influence of the dynamics between countries on determining gas prices, highlighting the intricate nature of gas trade and the role of bilateral relationships in price fluctuations (Fig. 5).

### 6.3   2020 - 2019 Network

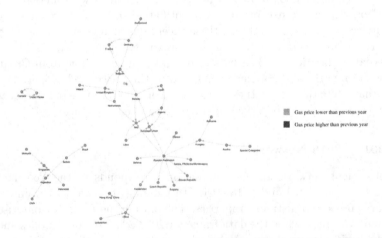

**Fig. 6.** Top 30% of the YoY network sorted by trade value for the years 2020-2019

The examination of this network holds significant interest as it allows us to observe the price dynamics surrounding a global pandemic. The focus lies on comparing price changes before and during the pandemic, shedding light on the effects of this unprecedented event. Specifically, we analyze the top 30% of natural gas trades to gain insights into the behavior of prices. In Fig. 6, a noteworthy observation is that all edges within this subset of trades are colored orange, indicating a noticeable decrease in price. From the total network, out of the 152 links, 131 experience a decrease in the price. This outcome aligns with expectations and can be attributed to the circumstances surrounding the pandemic. In 2019, during normal functioning conditions worldwide, there was an increase in demand for electricity and heating energy, leading to a corresponding rise in prices. However, in 2020, as most of the world implemented lockdown measures, major energy-consuming facilities were forced to shut down, resulting in a sharp decline in demand. According to the principles of supply and demand, an oversupply coupled with a substantial reduction in global demand leads to a decrease in the price of natural gas. Furthermore, when considering the nodes and edges before and after merging, consistent patterns are once again observed. This confirms the repeatability of the observed behavior, reinforcing the reliability and validity of the findings within the network analysis.

## 6.4    2021 - 2020 Network

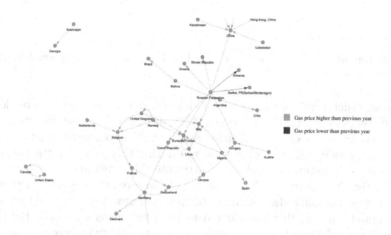

**Fig. 7.** Top 30% of the YoY network sorted by trade value for the years 2021-2020

In contrast to the previous network, characterized by a sudden and widespread lockdown, the current scenario reflects a gradual return to normalcy. This transition has resulted in dynamic shifts in people's lifestyles and various economic indicators, including the prices of stocks and energy commodities. These profound changes are likely to have influenced the patterns observed in the graph. Figure 7 reveals that the majority of links - 131 out of 145 in the total network are depicted in orange, indicating that countries have purchased natural gas at higher prices. This trend aligns with expectations, considering the historical context of energy commodities. In 2020, the price of energy commodities experienced an unprecedented decline, reaching an all-time low. Therefore, the observed behavior of countries purchasing natural gas at higher prices is consistent with the recovery of energy commodity prices from their previous lows.

## 6.5    2022 - 2021 Network

**Fig. 8.** Top 30% of the YoY network sorted by trade value for the years 2022-2021

The most crucial network analyzed in this research is the recent depiction of natural gas price fluctuations between 2022 and 2021. This network holds significant relevance for our investigation as it unveils important insights into the factors influencing the price of natural gas during this period. Anticipating an upward trend in natural gas prices, we attribute this development to several factors. Firstly, the impact of inflation on the global economy has exerted pressure on gas prices. Secondly, the scarcity of natural gas, as widely reported by various media outlets, has further contributed to the price increase. Lastly, the tightened political relations between partnering countries have emerged as the most influential factor driving the changing price dynamics. From Fig. 8, it becomes evident that the majority of edges are represented by an orange color - a record 141 of 153 total links - or a record 92%, indicating higher natural gas prices in 2022 compared to 2021. However, a few exceptions challenge this prevailing trend. Specifically, the partnerships between Russia and Serbia, as well as Russia and Armenia, exhibited a counter-intuitive pattern, where natural gas was purchased at even lower prices than the previous year. This exceptional finding suggests that in certain partnership cases, factors beyond the global price increase trend, such as political relations, play a crucial role in determining the purchasing price of natural gas.

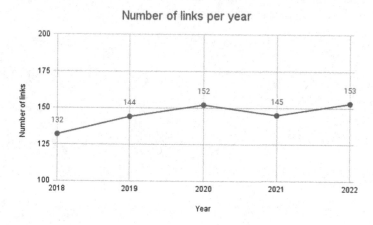

**Fig. 9.** Number of links per year

## 6.6    YoY Networks Conclusion

From Fig. 9, we observe the distribution of the number of links in the YoY networks per year. From 2018 to 2020, the trend line is increasing, indicating a more globalized trade and forming many new trading partnerships. However, from 2020 onward we see a slight decrease in the number of links, which might be influenced by the Covid pandemic and decreased overall trade of all commodities, including natural gas.

In our analysis of the YoY (Year-over-Year) graphs, several noteworthy trends emerge regarding the links between price changes and years. Firstly, during the period spanning from 2018 to 2020, a discernible decline is observed in the number of links exhibiting price increases. This decrease may be attributed to external factors, such as variations in weather conditions and the intrinsic abundance of the commodity in question. Subsequently, starting from 2020 and extending onward, a marked uptick is evident in the number of positive links. This surge in positive links might be directly attributed to the price decrease of natural gas and other energy commodities in 2020, which resulted from the global lockdown measures. Further analysis of the years 2021 and 2022 reveals a moderate increase in the number of higher-priced links. However, it is noteworthy that the proportion of all higher-priced links remains notably high in both cases, accounting for 90% and 92% respectively. This indicates a global trend wherein countries experienced elevated purchase prices for this particular commodity. Such an increase might be attributed to the Covid-related lockdowns and the emergence of political tensions in 2022 (Fig. 10).

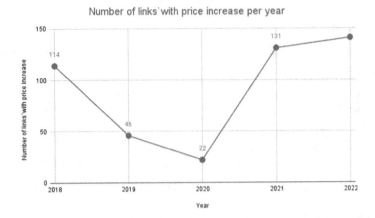

**Fig. 10.** Number of links with price increase per year

## 7   Discussion

The primary objective of this research is to examine the evolution of the gas trade over time. It is assumed that the price of natural gas adheres to the principles of supply and demand, similar to other comparable commodities. However, within the Year-over-Year (YoY) networks, there are two notable cases influenced by major global events. The first case pertains to the price discrepancy between 2020 and 2019, coinciding with the outbreak of the world pandemic. During this period, natural gas prices experienced a decline due to the closure of major energy-consuming facilities, resulting in a decrease in demand for natural gas. The second special case occurs in the 2022-2021 network, coinciding with the onset of the war in Ukraine. In this scenario, it is expected that prices would rise due to heightened political tensions, which is indeed observed. With the exception of only two key-value pairs where prices were lower, the network generally follows the trajectory of price increase attributable to higher demand and inflation. In summary, this research focuses on the analysis of how the gas trade has evolved over time. It is presumed that natural gas prices adhere to supply and demand dynamics, akin to other similar commodities. However, specific instances within the YoY networks exhibit deviations influenced by significant global events. The first case corresponds to the pandemic-induced decline in prices, while the second case involves price increases due to escalating political tensions in Ukraine.

## 8   Conclusion

Conclusions can be derived from each section of this research, shedding light on the role of Europe and its countries in the global gas trade and the evolving dynamics of interconnectivity within the region. The findings indicate a growing globalization of the gas trade, offering favorable prospects for countries to enhance the diversity of their natural gas import sources. The initial section of the study identifies the primary gas exporters, namely the Russian Federation, Norway, Algeria, Turkmenistan, and Belgium. Analyzing the in-degree centrality, the European Union, United Arab Emirates, and Bulgaria demonstrate the most diversified sources of gas, while the out-degree centrality reveals that the United States, the Russian Federation, and the United Kingdom export to a wide range of countries. Furthermore, the eigenvector centrality measure identifies the United States, Mexico, and Canada as countries of high importance in the gas trade. Regarding the volume of natural gas import and export, Germany, the EU, and Italy emerge as the largest importers, while the Russian Federation, Special categories, and Norway rank highest in terms of export quantity. Spain, Italy, and France are identified as key nodes with advantageous positions within the gas trade networks. The final section, focusing on the YoY graphs, provides insights into the factors influencing natural gas prices. The analysis highlights the significant impact of supply and demand dynamics, as well as inflation, on gas prices. Additionally, the geographical location of partner countries and their political relations are found to be influential factors. During the COVID-19 pandemic, gas prices exhibited a similar trajectory across all countries, coinciding with a global increase in inflation rates. This suggests that transformative world events play a substantial role in shaping gas prices alongside the fundamental forces of supply and demand. However, certain exceptional cases deviated from this global trend, indicating the presence of additional factors affecting natural gas prices.

# References

1. Bihari, A., Pandia, M.: Eigenvector centrality and its application in research professionals' relationship network. In: 2015 1st International Conference on Futuristic Trends in Computational Analysis and Knowledge Management, ABLAZE 2015 (2015). https://doi.org/10.1109/ABLAZE.2015.7154915
2. Belgium Ministry of Energy: Ministry of energy, Belgium: Natural gas policy (2022). https://www.energy.belgium.be/national-natural-gas-policy-2022
3. Filimonova, I.V., Komarova, A.V., Sharma, R., Novikov, A.Y.: Transformation of international liquefied natural gas markets: new trade routes. Energy Rep. **8** (2022). https://doi.org/10.1016/j.egyr.2022.07.069
4. Geng, J.B., Ji, Q., Fan, Y.: A dynamic analysis on global natural gas trade network. Appl. Energy **132** (2014)
5. Grigas, A.: The New Geopolitics of Natural Gas. Harvard University Press (2017)
6. Halser, C., Paraschiv, F.: Pathways to overcoming natural gas dependency on Russia - the German case. Energies **15**, 4939 (2022). https://doi.org/10.3390/en15144939
7. Lu, H., Ma, X., Ma, M., Zhu, S.: Energy price prediction using data-driven models: a decade review. Comput. Sci. Rev. (2021). https://doi.org/10.1016/j.cosrev.2020.100356
8. Jana, R.K., Ghosh, I.: A residual driven ensemble machine learning approach for forecasting natural gas prices: analyses for pre-and during-covid-19 phases. Ann. Oper. Res. (2021). https://doi.org/10.1007/s10479-021-04492-4

9. Li, J., Dong, X., Jiang, Q., Dong, K., Liu, G.: Natural gas trade network of countries and regions along the belt and road: where to go in the future? Resources Policy (2021). https://doi.org/10.1016/j.resourpol.2020.101981
10. Zhang, J., Luo, Y.: Degree centrality, betweenness centrality, and closeness centrality in social network (2017)
11. Klein, D.J.: Centrality measures in graphs. J. Math. Chem. (2010). https://doi.org/10.1007/s10910-009-9635-0
12. Lin, N., Brooks, R.E.: Global liquified natural gas trade under energy transition. Energies **2021** (2021). https://doi.org/10.3390/en14206617
13. Norouzi, N.: Post-covid-19 and globalization of oil and natural gas trade: challenges, opportunities, lessons, regulations, and strategies. Int. J. Energy Res. (2021). https://doi.org/10.1002/er.6762
14. Newman, M.E.J.: The structure and function of complex networks. SIAM Rev. **45**, 167–256 (2003). https://doi.org/10.1137/S003614450342480
15. Opsahl, T., Agneessens, F., Skvoretz, J.: Node centrality in weighted networks: generalizing degree and shortest paths. Soc. Netw. **32**(3) (2010). https://doi.org/10.1016/j.socnet.2010.03.006
16. Ebrahimpour-komleh, H., Lazemi, S.: Computing connected components of graphs. Int. J. Appl. Math. Res. (2014)
17. Kan, S.Y., Chen, B., Wu, X.F., Chen, Z.M., Chen, G.Q.: Natural gas overview for world economy: from primary supply to final demand via global supply chains. Energy Policy (2019). https://doi.org/10.1016/j.enpol.2018.10.002
18. Trajanov, D., Vodenska, I., Cvetanov, G., Chitkushev, L.: Data driven analysis of trade, FDI and international relations on global scale. In: The 13th Annual International Conference on Computer Science and Education in Computer Science (2017)
19. UN: United nations comtrade database (2022). https://comtrade.un.org/
20. Schulhof, V., van Vuuren, D., Kirchherr, J.: The belt and road initiative (BRI): what will it look like in the future? Technol. Forecasting Soc. Change (2022). https://doi.org/10.1016/j.techfore.2021.121306
21. Vodenska, I., Trajanov, D., Trajanovska, I., Chitkushev, L.: Impact of global events on crude oil prices. In: 12th Annual Research Conference on Computer Science and Education in Computer Science, Fulda and Nürnberg, Germany (2016)

# Identifying Drug - Disease Interactions Through Link Prediction in Heterogeneous Graphs

Milena Trajanoska[✉], Martina Toshevska, and Sonja Gievska

Faculty of Computer Science and Engineering, Ss Cyril and Methodiuos University, Skopje, North Macedonia
{milena.trajanoska,martina.toshevska,sonja.gievska}@finki.ukim.mk

**Abstract.** Unlike traditional development of new drugs that rely on labor- and time-intensive research and clinical trials, computational approaches, deep learning technologies, in particular, have been prominent in recent research on the topic. By utilizing the ever-growing biomedical knowledge repositories and exploiting the relationship between diverse types of information (e.g., proteins, genes, molecular, diseases, drugs), graph neural networks (GNNs) primed for processing graph-structured data have a real potential for advancing the critical endeavor of drug discovery. Safe and effective drug therapy would also rely on early identification of unwanted and potentially harmful adverse effects a certain drug has on patient's health. Hence, two, rather contrastive tasks that pertain to the process of drug discovery have been of special interest in this research. The first one is drug repurposing and the second one, a closely-related task of identifying drugs that have an adverse or negative effect on patient health namely drug-induced diseases. In this research, the task of discovering new links between drugs and diseases has been formalized as a link prediction task in a heterogenous graph. The predictive models for drug discovery proposed in this paper were tested on the ogbl-biokg (https://ogb.stanford.edu/docs/linkprop/#ogbl-biokg) dataset from the collection of large benchmark dataset Open Graph Benchmark (OGB) [15]. The openness and multi-source heterogeneity of the OGB dataset has provided us with an opportunity to experiment with HinSage [28], a method for inductive representational learning in heterogenous graphs. Two models based on HinSage, have been proposed proving their superior performance when compared with more traditional similarity-based baseline methods. Furthermore, a selected newly discovered relationship with a potential for drug repurposing has been discussed through the lenses of related clinical-experimental trials.

**Keywords:** Discovering Drug-Disease Interactions · Heterogeneous Graph · HinSAGE · Link Prediction · Contextual Enrichment

Supported by Faculty of Computer Science and Engineering, Skopje, N. Macedonia.

M. Mihova and M. Jovanov (Eds.): ICT Innovations 2023, CCIS 1991, pp. 191–205, 2024.
https://doi.org/10.1007/978-3-031-54321-0_13

# 1  Introduction

The process of new drug discovery, also called *de novo* drug discovery, is a laborious and financially demanding procedure characterized with a notable failure rate, as the majority of potential drug candidates are excluded from further consideration during their clinical trials. The risk factors associated with the process include the use of untested chemical compounds, high research and development costs, and extensive clinical trials.

For this reason, drug repurposing, also known as drug repositioning, is becoming a popular alternative. It is described as the process of identifying new therapeutic uses for existing drugs that are already approved, or in clinical development for other indications. The popularity of this approach comes from the lowered risks, costs, and time period for developing a drug for a certain disease of interest.

In general, there are two approaches to drug repurposing, namely experimental and computational [22]. The experimental approaches include: binding assays to measure relevant interactions, as well as phenotypic screening. Before the advancements in the fields of deep learning and graph-based computation, the algorithmic methods included more traditional techniques, such as: signature matching, pathway mapping, genetic association, molecular docking, and retrospective clinical analysis [22].

However, these recent advancements in deep learning have accelerated the research process and enabled using a combination of topological information of the drug-disease interactions along with their structural (or compound-based) information. Moreover, the potential for drug-induced diseases, along with side effects, often remains hidden during clinical trials of a drug. This concern is critical as rushed testing and incomplete assessments could result in adverse effects for patients. For instance, Thalidomide, once used to alleviate nausea in pregnant women, caused severe birth defects in thousands of newborns later on [24]. Therefore, utilizing computational approaches to predict potential drug-induced diseases has the potential to enhance the effectiveness of drug trials significantly.

The onset of our study, a heterogeneous graph, consisting of nodes, representing drugs and diseases, was constructed. Two types of links have been defined, namely drug-disease links and drug-drug links. A link between a drug and a disease exists, if the drug is known to treat or to cause a certain disease. A link between two drug nodes (drug-drug link) represent a known interactions between a pair of drugs when administered in a combined drug treatment.

The problem of discovering new drug-disease links was defined as a link prediction task in an incomplete knowledge graph. The knowledge graph is incomplete because not all existing drug-disease relations have been discovered or experimentally confirmed.

In our study, multiple methods for link prediction have been empirically investigated and their performance results have been compared. Each newly discovered drug-disease relationship could be further explored for its potential benefit in drug repurposing or its potential negative imapact on patients' health (drug-induced diseases). A number of similarity-based methods, such as: Jaccard Coefficient [21], Preferential Attachment [20], and Adamic Adar [2] have been used as baseline methods. In addition, three predictive graph-based models have been considered: 1) A model that uses Metapath node embeddings [11] and a Random Forest [6] classification algorithm, 2) a graph neural network, HinSAGE, which is a heterogeneous graph version of the algorithm GraphSAGE [13], and 3) an augmented version of the HinSage model using additional drug features as node embeddings.

All methods were evaluated using a subset of the ogbl-biokg knowledge graph dataset. This subset focused on two entity types, drugs and diseases, and encompassed drug-drug and drug-disease relationships. The resulting heterogeneous graph included 10,533 drugs and 10,687 diseases, featuring a total of 1,138,833 edges. Notably, 5,147 of these edges represented drug-disease interactions.

The results have confirmed the performance superiority of both HinSAGE-based models achieving 0.93 ROC AUC and 0.92 ROC AUC, respectively, on the testing dataset. Finally, a set of the newly identified drug-disease interactions has been discussed in relation to our predictions.

This paper is organized as follows: the next section summarizes the research on the topic closely related to our own. The third section presents detailed description of the dataset used for evaluating the performance of the proposed models and the specifics of the evaluated predictive models. The forth section, includes a discussion of the experimental results obtained by the models evaluated in this study and a brief interpretation of few newly discovered drug-disease interactions. The paper concludes with an overview of the limitations and practical implications of the results presented in this research.

## 2   Related Work

Amidst the ongoing advancements in network computations and graph neural networks (GNNs), considerable efforts have been made to apply these methods in diverse domains, such as social network analysis, recommendation systems, biological network analysis, traffic prediction, and language modeling tasks. The accelerating potential of graph representation learning in discovery of previously unexplored drug-disease interactions is evident in recent works [1,4,8,14,16,18, 26,31–36]. In what follows, we discuss the recent research efforts dedicated to the study of drug repurposing.

A multi-labeled drug repurposing method called DR-HGNN has been proposed by Sadeghi et al. [26]. A heterogeneous graph, consisted of 708 drug nodes and 1,512 protein nodes, and 39 diseases representing 5,603 disease interactions was constructed by extracting information from DTINet dataset[1]. An edge labeled with a disease name, was associated with each a pair of (drug, protein) nodes, if a disease is associated with both the drug and protein being examined. The problem of drug-disease prediction was defined as a multi-label prediction

---

[1] https://github.com/luoyunan/DTINet.

task, The DR-HGNN model has obtained an ROC AUC score of 0.96 on the DTINet dataset.

Zhao et al. [36], use an algorithm named Multi-Graph Representation Learning (MGRL) for predicting drug-disease associations. The nodes representing drugs have been augmented with drug chemical characteristics in the form of a simplified specification from the Molecular Input Line Entry System (SMILES). Semantic description of diseases was used as additional information in nodes representing diseases. node2vec embedding algorithm was used to learn representations of the drugs-disease relationships. The model has achieved ROC AUC score of 0.85 using a Random Forest classifier on the Comparative Toxicogenomics Database (CTD) dataset[2], from which they have curated a total of 18,416 drug-disease pairs, between 269 drugs and 598 diseases.

Yu et al. [33] introduced a computational approach named layer attention graph convolutional network (LAGCN) intended for drug-disease prediction, which they have evaluated on the CTD dataset. The heterogeneous graph integrated previously identified drug-disease interactions, drug-drug similarities, and disease-disease similarities. It was created to learn their representation and predict unobserved drug- disease associations. The reported performance of the model was an area under the ROC curve of 0.8750.

A method called DRHGCN - Drug Repositioning based on the Heterogeneous information fusion Graph Convolutional Network, with the goal of discovering potential drugs for a certain disease [8]. Several networks have been utilized, namely, drug-drug similarity, disease-disease similarity as well as drug-disease association networks. Attention mechanism was used to identify the most relevant inter- and intra-domain features extracted using graph convolutional operations to the networks. The model has been evaluated on four benchmark datasets achieving an average area under the ROC curve of 0.934.

A bipartite graph convolution network model intended for drug repurposing, using heterogeneous information fusion, called BiFusion was proposed by Wang et al. [31]. The graph includes drug-disease, drug-protein, protein-protein, and disease-protein interactions, and was generated to compile the insights of multiscale pharmaceutical information. Protein nodes were introduced as bridges for message passing across biological domains. The model has achieved a 0.857 area under the ROC curve score using a 10-fold cross-validation on the repoDB dataset[3]. The dataset contains 592 disease, 1,012 drug and 13,460 protein nodes. The nodes are connected by 3,204 drug-disease, 104,716 disease-protein, 7,713 drug-protein, and 141,296 protein-protein edges.

Han et al. [14] describe SmileGNN, an approach that combines topological information derived using GNNs from drug-disease interaction networks, with structural information derived from SMILES sequences. The authors compare their solution with existing algorithms on two datasets, The Kyoto Encyclopedia

---

[2] https://ctdbase.org/.

[3] https://repodb.net/.

of Genes and Genomes[4], and the Patient, Disease, and Drug dataset[5], containing 56,983 and 36,768 drug-disease edges, respectively. The ROC AUC score of the SmileGNN on these datasets was 0.9521 and 0.9642, respectively.

An approach utilizing a heterogeneous information network (HINGRL) [35] has been implemented for predicting potential drug-disease association using graph representation learning techniques. For representing the features of drugs and diseases, HINGRL constructs complex heterogeneous information network (HIN) which contains integrated protein-related associations with biological knowledge of drugs and diseases. The authors show that by incorporating biological knowledge, the performance of the model was improved. The prediction task is performed using the Random Forest classifier on embeddings generated from the HIN. A benchmark dataset comprised of a combination of the CTD, the DrugBank[6], and the DisGeNET[7] database is used for training and evaluation.

DeepDR is an approach for in silico drug repurposing [34] that integrates data from 10 biological networks to capture relationships between drugs, diseases, side effects, and interactions. It employs a multi-modal deep auto-encoder to extract key drug features from these networks. The model was trained on a graph containing 6,677 known drug-disease pairs, achieving a high performance with an ROC curve area of 0.908. Validation on 129 new drug-disease pairs confirmed its effectiveness.

A more recent study [16] proposes an inductive recurrent graph convolutional network (RGCN) for learning relation embeddings in a few-shot learning setting. The algorithm uses a multi-layer perceptron (MLP) to learn the embedding of the desired relation by using only a few examples. The authors apply the method on the drug-repurposing knowledge graph (DRKG) with the goal of discovering drugs for Covid-19.

Jin et al. have proposed a drug repositioning framework which is based on heterogeneous information networks and algorithms from text mining called HeTDR [18]. The model extracts drug feature representations from multiple drug-related networks. In additional, disease features are constructed from biomedical corpora. The network is created using known drug-disease associations with the goal of predicting new drug-disease associations. The features extraction for the drugs is based on similarity network fusion (SNF) [30] and sparse auto-encoder (SAE). The vector representation of a disease is learned from biomedical corpora using a transformer model for the biomedical domain, namely BioBERT [19].

Because the process of identifying drug-disease interactions has a large impact on administering a combined drug therapy, it is necessary to explain the results of drug-disease association algorithms in order to justify their existence. He et al. [1] have implemented an explainable model for drug repositioning abbreviated EDEN. EDEN fuses local and global semantics from a disease interaction network by maximizing mutual information. Using the learnt embeddings, the

---

[4] https://www.genome.jp/kegg/.

[5] https://github.com/wangmengsd/pdd-graph.

[6] https://go.drugbank.com/.

[7] https://www.disgenet.org/.

method contributes to the explainability of the results by calculating global paths between the predicted associations. The empirical results presented in the study have shown that EDEN outperforms multiple state-of-the-art baselines when evaluated on a variety of benchmark datasets including CTD, repoDB and DrugBank.

We would like to highlight two major aspects distinguishing our approach from the research endeavors closely related to our task at hand. The main one being that we use an end-to-end graph representation method HinSage [28] to learn and predict drug-disease associations directly from the constructed heterogenous graph, as opposed to independent learning of drug and disease embeddings either from multiple networks or a heterogeneous graph as adopted in other studies. The other fact worth mentioning is the greater number of drug and disease nodes included in the heterogenous graph used in our study, with a notation that a comparison between the evaluation results of related research is not a straightforward task fur to the variance of information on which graphs are constructed.

## 3    Methods

Our tasks under investigation, namely: drug repurposing and drug-induced diseases are closely related to the task of link prediction of new drug-disease interactions in a heterogenous graph containing information on drugs and diseases. In what follows, we describe the dataset used in the evaluation study, the construction of the heterogenous graph and the proposed prediction models.

### 3.1    Dataset

Comparative evaluation of the performance of the proposed models were conducted on a subset of the Open Graph Benchmark ogbl-biokg[8] dataset [15], that represents a Knowledge Graph that has been curated utilizing information sourced from an extensive array of biomedical data repositories. This dataset encompasses entities categorized into five distinct types: diseases (10,687 nodes), drugs (10,533 nodes), side effects (9,969 nodes), proteins (17,499 nodes), and protein functions (45,085 nodes). There are 51 types of directed relations connecting entities of the same or different type, including 38 kinds of drug-drug interactions, 8 kinds of protein-protein interaction, as well as drug-protein, drug-disease, drug-side effect, and protein-protein function relations.

For the purpose of this study, two types of entities have been extracted from the entire dataset, namely disease and drug nodes. In addition, the entire set of 38 types of drug-drug interaction edges and the edges representing drug-disease interaction have been included in a heterogeneous network, with a total of 21,220 nodes and 1,138,833 edges, out of which 1,133,686 were drug-drug interactions and 5,147 were drug-disease interactions.

---

[8] https://ogb.stanford.edu/docs/linkprop/#ogbl-biokg.

## 3.2   Training, Validation and Test Dataset

We measured the performance of the proposed link prediction models using cross-validation, with 70% to 10% to 20% separation of the entire set of edges. The training, validation and test subsets of edges contain both positive (previously confirmed drug-disease interactions) and negative (unknown/nonexistent drug-disease interactions) edges. The ratio of positive and negative edges in both the training and validation subsets is maintained at 1:1. However, in the testing subset, the ratio of negative to positive edges is approximately 32:1, which deems reasonable for the tasks at hand, which situations involving prediction of unfamiliar drug-disease interactions, particularly in tasks like drug repurposing or identifying new drug-induced diseases.

The procedure for sampling negative edges was carried out as follows: for each drug in the graph, a maximum of 10 diseases with which the drug does not have a known interaction were sampled at random. This procedure resulted with the edge splits shown in Table 1.

**Table 1. Number of positive and negative edges per split.** The rows represent the subset of data and the columns represent the number of positive and negative edges respectively.

|                    | Positive edges | Negative edges |
| ------------------ | -------------- | -------------- |
| Training subset    | 2,574          | 2,574          |
| Validation subset  | 515            | 515            |
| Testing subset     | 2,060          | 65,513         |

## 3.3   Graph Construction

A heterogenous graph consisting of 21,220 nodes and 1,138,833 edges was created; the undirected graph edges represent 1,133,686 different types of drug-drug interactions and 5,147 drug-disease interactions.

The generated graph has 10,462 connected components, signifying the presence of 10,462 sets of nodes within the graph between which a path of edges is absent (not connected). The largest connected component encompasses 10,741 nodes, accounting for approximately half of the total nodes constituting the heterogeneous graph. Most of the nodes in the network (10,450) have a node degree of 0. A total of 9,172 nodes with degree 0 represent the diseases which has no previously known relation/interaction with any other disease, nor is there an approved drug to cure these diseases. The other 1278 nodes with degree 0 are drug nodes that are not connected to a disease, meaning these drugs exist but they are not used as a treatment to any of the diseases, nor do they have any known interactions with other drugs present in the current dataset. The sparsity of the graph poses an additional challenge for correctly identifying previously unknown drug-disease interactions.

## 3.4   Similarity-Based Link Prediction Methods

At the onset of our experimental study, three baseline similarity-based link prediction models were established against which the performance of the proposed graph-based neural models were compared. The selected similarity metrics on which these baseline models were based upon were the well-know: Adamic Adar [2], Jaccard Coefficient [21] and Preferential Attachment [20] and their description follows. A similarity coefficient was calculated for each pair of nodes to obtain values which quantifies the existence of a relationship between the nodes.

The Adamic Adar similarity metric, introduced in their study on prediction link structure based on information pertaining to social interaction [2]. By taking the premise that the nodes shared by large neighborhoods are less significant than the nodes shared by neighborhoods with small number of nodes, Adamic Adar similarity is defined as follows:

$$A(x,y) = \sum_{u \in N(x) \cap N(y)} \frac{1}{\log|N(u)|} \tag{1}$$

where A(x, y) represents the similarity between nodes x and y, N(x) is the neighborhood of node x and $|N(x)|$ represents the number of nodes in the neighborhood of x.

The Jaccard Coefficient [21] is defined in the following way:

$$J(x,y) = \frac{|N(x) \cap N(y)|}{|N(x) \cup N(y)|} \tag{2}$$

where the similarity is measured as the ratio between the number of overlapping (shared) nodes in the neighborhoods of x and y and the total number of nodes in their union.

The third baseline model we have used is based on the similarity metric called Preferential Attachment [20], computed using Eq. 3.

$$P(x,y) = |N(x)| * |N(y)| \tag{3}$$

The underlying premise of Preferential Attachment is that nodes with higher degree of connectivity are inclined to form additional connections with other nodes.

A threshold value for each of the similarity metrics was experimentally selected, namely: 0.15, 0.012 and 12,000 for the Adamic Adar, Jaccard coefficient and Preferential Attachment, respectively. The nodes with values above the threshold were classified as having a link between them, while the nodes with similarity value below the threshold were assumed to have links between them. The cutoff threshold was experimentally chosen with the rationale that the nodes which are high similar will belong to the long tail of the distributions, meaning that they would have similarity value significantly larger than 0.

The distributions of the similarities between pairs of nodes in the testing dataset are displayed in Fig. 1.

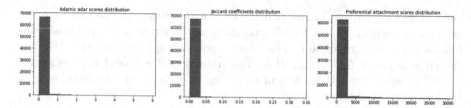

**Fig. 1. Distribution of the node pair similarities calculated for the testing dataset.** The three subfigures correspond to the distribution of the calculated Adamic Adar, Jaccard coefficients and the Preferential Attachment, respectively.

### 3.5  Link-Prediction Using Metapath2vec Node Embeddings

In predictive modeling on graph-based data, a number of representational learning methods have been proposed [10]. We have selected two of them that are applicable to heterogenous graphs, one of them being the metapath2vec [11] algorithm. In brief, the metapath2vec implements a heterogeneous skip-gram model to create node embeddings in a heterogenous graph by executing metapath-based random walks. A metapath refers to the path i.e. types of links traversed between different node types when generating a node embedding. For our link prediction task, the following metapaths have been utilized:

- $drug \rightarrow disease \rightarrow drug$
- $drug \rightarrow disease \rightarrow drug \rightarrow drug$
- $drug \rightarrow drug$

Ten random walks were generated per node, each of length 5. For each link in the training subset, the node embeddings were concatenated and passed as a feature vector to a Random Forest Classifier [6] that performs the link prediction task. The validation subset was used for hyper-parameter tuning, the best performing model was evaluated on the testing subset.

### 3.6  Link Prediction with HinSAGE

In this study, we propose an end-to-end approach for graph representation learning in a heterogeneous graph. In particular, we have evaluated the performance of the generalized version of GraphSAGE [13], called HinSage [28], which is primed for representation learning in graphs containing multiple types of nodes and edges. To account for the heterogeneous information, multiple weight matrices are added.

To address the problem posed in this study, a two-layered HinSAGE model was used to predict links between drug-disease node pairs. The optimal number of neighbors in each layer was experimentally set to 8 by performing multiple hyper-parameter tuning steps on the validation dataset. On top of the two-layer HinSAGE model, a link embedding layer was added which produces link embeddings by concatenating the node embedding for a drug with the node embedding for a disease. Finally, a dense layer with a single neuron and a sigmoid activation function [23] set to 0.5 was added as the model head, to classify the link as being present or not. The binary cross-entropy loss function [17] was chosen as a target for optimization. The training of the model was performed for 100 epochs with a batch size of 64, using an early stopping conditioned on the loss on the validation subset.

### 3.7    Drug-Feature Augmented HinSage

Augmenting node and edge representations with features that reflect the intrinsic properties of the types of entities and relationships they represent is frequently pointed to improve the classification accuracy. By doing this, node and edge embeddings not only encode the underlying network structure, but also carry the information, indicators and properties of the specific entity or relationship.

In our setting, the HinSAGE architecture described in the previous subsection was strengthen with drug-related features. The information related to drugs was obtained from the PubChem NCBI database [29], which included chemical characteristics, such as: molecular weight, polar area, complexity, octanol/water partition coefficient calculated by the atom-additive method xLogP, heavy compounds count, and weather the main compound is a hydrogen bond donor. For drugs with no available information in the PubChem NCBI database, a vector of ones with dimension equal to six was added as a feature vector. The inclusion of drug-related information was expected to help the model differentiate between drugs on the basis of their specific chemical properties, as well as their neighborhood structural similarities. Ultimately, drugs with similar chemical and structural properties should be close in the vector space.

## 4    Discussion of Results

This section describes the experimental results of our exploratory study as well as some insights into the prediction of potential drug-disease interactions which might represent candidates for further investigation on the tasks of drug repurposing and discovering drug-induced diseases.

For predicting drug-disease interactions, a machine learning model which would be able to discover valid, yet unknown drug-disease interactions accurately, is needed. A suitable evaluation metric, the area under the ROC curve [5], abbreviated AUC was chosen. The metric represents the ratio between the generated true positives and false positives and gives a much better insight into the strengths and weakness of the predictive models. Recall metric, which is calculated as the fraction of accurately predicted positive samples against the actual number of positive samples in the dataset was used to quantify how good the models classify the previously confirmed (positive) drug-disease relations.

The Area under the ROC curve is optimized while evaluating the model on the validation subset, for reference the recall score is also reported. The results from the models for link prediction on the testing subset, described in the methods section are displayed in Table 2. The results show the superiority of the three graph-based models when compared against the performance of the similarity-based baseline models, which was expected.

However, the performance comparison of the three graph-based learning methods highlights the subtle similarities and differences between their underlying algorithms. Having in mind that random walks along different types of meta-paths is at the core of both representational learning, the metapath2vec and HinSAGE, no notable difference on the ROC AUC values was found. A small and yet significant improvement in the recall score exhibited by the original and augmented version of HinSAGE is evident when compared to the metapath2vec-based model. It should be noted that Positively classifying existing drug-disease links is crucial for some application problems, thus metrics such as recall quantifying the advantages or failings in some models should be taken into account.

**Table 2. Experimental results.** The table represents the calculated ROC AUC score and recall score on the test subset.

| Algorithm | Testing ROC AUC score | Testing Recall Score |
|---|---|---|
| Adamic Adar | 0.76 | 0.53 |
| Jaccard Coefficient | 0.66 | 0.33 |
| Preferential Attachment | 0.73 | 0.48 |
| metapath2vec | 0.92 | 0.97 |
| HinSAGE | 0.92 | 1.0 |
| Drug-feature augmented HinSAGE | 0.93 | 0.99 |

The interpretation of the results obtained by the original HinSAGE model and the augmented HinSAGE model is not a straightforward task. A look at the confusion matrices of the six models we have evaluated shown in Fig. 2 might help in understanding the differences and hypothesized about the reasons underlying them. It should be noted that the original HinSAGE algorithm correctly classifies all positive edges, while the augmented HinSAGE model has only 4 false negatives. However, the drug-feature augmented HinSAGE model has 393 less false positives, compared to the original HinSAGE model. Different application settings may afford different models and one might argue that the subtle differences in practical real-world uses might require a unique modelling approach.

The ROC AUC advantage of the augmented HinSAGE model might be attributed to the fact that adding chemical characteristics of a drug's main active compound might be helpful in filtering drug-disease relationships that are less likely to be significant. The small and yet significant positive performance change are encouraging and yet point out to future research direction that might be fruitful. It would be interesting to analyze and compare the impact of augmenting both types of nodes, drugs ad diseases, with additional features.

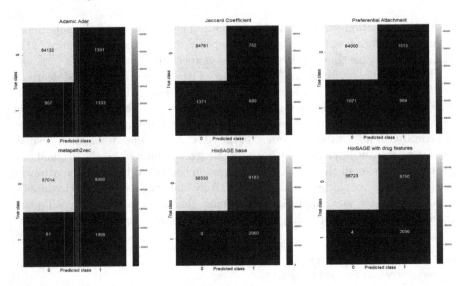

**Fig. 2.** Confusion matrices of the six models for predicting new drug-disease links on the testing dataset.

## 4.1  Case Study

This subsection evaluates predicted links between drugs and diseases, which were previously unknown, using the predictions generated by the Drug-feature augmented HinSAGE model due to its demonstrated reliability in producing results.

One of the interesting relations found is connected to the drug Blenoxane. Blenoxane, with generic name bleomycin sulfate injection, is an Antineoplastics, Antibiotic medication employed for alleviating the manifestations of Squamous Cell Carcinoma, Non-Hodgkin Lymphoma, Testicular Carcinoma, Hodgkin Disease, and Pleural Sclerosing [25]. In our case, the model additionally predicts that the drug has relations with cervical cancer, penile cancer, posterior urethra cancer, and thyroiditis.

A study [7] indicates that a combination of belomycin, ifosfamide, and cisplatin demonstrates potent efficacy in cases of advanced and recurrent cervical cancer. Additionally, other studies exist [3,12] where experiments were conducted using bleomycin with a combination of other medications in chemotherapy for penile cancer. However, in both cancer cases, this drug has still no proven effects. Moreover, one study [27] presents a rare instance of a 59-year-old female patient manifested autoimmune-mediated hepatitis and Hashimoto's thyroiditis during the preliminary diagnosis of Hodgkin lymphoma. The woman underwent treatment with bleomycin which resulted in a complete remission. This goes in hand with our prediction that Blenoxane (bleomycin) has an interaction with thyroiditis.

On the other hand, Azathioprine an immunosuppressant, is used for treating rheumatoid arthritis, severe inflammation of the liver, Crohn's disease and ulcerative colitis, skin or arteries, as well as blood disorders. The model predicts that Azathioprine has an interaction with atrial fibrillation. One study [9] shows that a 52-year-old male patient manifesting steroid-dependent ulcerative colitis, experienced atrial fibrilation after taking Azathioprine as treatment. This shows that the model can predict both positive (treatment, cure) and negative (drug-induced diseases) outcomes of the relations between drugs and diseases.

## 5    Conclusion

Overall, this study represented a computational approach for drug-disease interactions prediction. Multiple experimental methods for link prediction were tested and their results were compared. The results showed that the HinSage-based models make the most accurate predictions of unknown drug-diseases interactions. Finally, some examples of predicted drug-disease interactions were shown and it was highlighted that there exist experimental research attempting to prove these interactions.

Ultimately, the study was limited due to the fact that the model was trained on an incomplete knowledge graph. Additionally, the interactions between drugs and diseases were not labeled as drug-cures-disease or drug-induces-disease, hence the model was trained to identify both types of interactions simultaneously, without differentiating between them. The differentiation was made during the phase of interpreting the results.

The implications of this study are significant on a practical level. Our entire pipeline can be reused with the task of predicting new drug-disease interactions in the real-world and can incrementally be improved by adding new discovered relations between drugs and diseases or between drugs.

Future steps in this research would include separating the drug-cures-disease and drug-induces-disease interactions, adding more entity and relation types to the heterogeneous graph such as proteins, functions, etc. Additionally, enriching the nodes with more features, using a similar procedure as the one for adding chemical characteristics as features of the drug nodes, as well as utilizing natural language descriptions of the corresponding entities is hypothesized to further improve the prediction results.

**Acknowledgements.** This work was partially financed by the Faculty of Computer Science and Engineering at the Ss. Cyril and Methodius University in Skopje.

## References

1. An explainable framework for drug repositioning from disease information network. Neurocomputing **511**, 247–258 (2022)
2. Adamic, L.A., Adar, E.: Friends and neighbors on the web. Soc. Netw. **25**(3), 211–230 (2003)

3. Ahmed, T., Sklaroff, R., Yagoda, A.: Sequential trials of methotrexate, cisplatin and bleomycin for penile cancer. J. Urol. **132**(3), 465–468 (1984)

4. Al-Rabeah, M.H., Lakizadeh, A.: Prediction of drug-drug interaction events using graph neural networks based feature extraction. Sci. Rep. **12**(1) (2022)

5. Bradley, A.P.: The use of the area under the ROC curve in the evaluation of machine learning algorithms. Pattern Recogn. **30**(7), 1145–1159 (1997)

6. Breiman, L.: Random forests. Mach. Learn. **45**(1), 5–32 (2001)

7. Buxton, E., et al.: Combination bleomycin, ifosfamide, and cisplatin chemotherapy in cervical cancer. JNCI: J. Natl. Cancer Inst. **81**(5), 359–361 (1989)

8. Cai, L., et al.: Drug repositioning based on the heterogeneous information fusion graph convolutional network. Briefings Bioinform. **22**(6), bbab319 (2021)

9. Cassinotti, A., et al.: New onset of atrial fibrillation after introduction of azathioprine in ulcerative colitis: case report and review of the literature. Eur. J. Clin. Pharmacol. **63**(9), 875–878 (2007)

10. Chen, F., Wang, Y.C., Wang, B., Kuo, C.C.J.: Graph representation learning: a survey. APSIPA Trans. Signal Inf. Process. **9**, e15 (2020)

11. Dong, Y., Chawla, N.V., Swami, A.: metapath2vec: scalable representation learning for heterogeneous networks. In: Proceedings of the 23rd ACM SIGKDD International Conference on Knowledge Discovery and Data Mining (2017)

12. Hakenberg, O.W., Nippgen, J.B., Froehner, M., Zastrow, S., Wirth, M.P.: Cisplatin, methotrexate and bleomycin for treating advanced penile carcinoma. BJU Int. **98**(6), 1225–1227 (2006)

13. Hamilton, W., Ying, Z., Leskovec, J.: Inductive representation learning on large graphs. In: Advances in Neural Information Processing Systems, vol. 30 (2017)

14. Han, X., Xie, R., Li, X., Li, J.: SmileGNN: drug-drug interaction prediction based on the smiles and graph neural network. Life **12**(2), 319 (2022)

15. Hu, W., et al.: Open graph benchmark: datasets for machine learning on graphs. In: Advances in Neural Information Processing Systems, vol. 33, pp. 22118–22133 (2020)

16. Ioannidis, V.N., Zheng, D., Karypis, G.: Few-shot link prediction via graph neural networks for Covid-19 drug-repurposing. arXiv preprint arXiv:2007.10261 (2020)

17. Janocha, K., Czarnecki, W.M.: On loss functions for deep neural networks in classification. arXiv preprint arXiv:1702.05659 (2017)

18. Jin, S., et al.: HeTDR: drug repositioning based on heterogeneous networks and text mining. Patterns **2**(8), 100307 (2021)

19. Lee, J., et al.: BioBERT: a pre-trained biomedical language representation model for biomedical text mining. Bioinformatics **36**(4), 1234–1240 (2020)

20. Newman, M.E.: Clustering and preferential attachment in growing networks. Phys. Rev. E **64**(2), 025102 (2001)

21. Niwattanakul, S., Singthongchai, J., Naenudorn, E., Wanapu, S.: Using of Jaccard coefficient for keywords similarity. In: Proceedings of the International Multiconference of Engineers and Computer Scientists, vol. 1, pp. 380–384 (2013)

22. Pushpakom, S., et al.: Drug repurposing: progress, challenges and recommendations. Nat. Rev. Drug Discov. **18**(1) (2019)

23. Ramachandran, P., Zoph, B., Le, Q.V.: Searching for activation functions. arXiv preprint arXiv:1710.05941 (2017)

24. Ridings, J.E.: The thalidomide disaster, lessons from the past. In: Barrow, P. (ed.) Teratogenicity Testing. MIMB, vol. 947, pp. 575–586. Springer, Cham (2013). https://doi.org/10.1007/978-1-62703-131-8_36

25. RxList: Blenoxane (2021). https://www.rxlist.com/blenoxane-drug.htm

26. Sadeghi, S., Lu, J., Ngom, A.: An integrative heterogeneous graph neural network-based method for multi-labeled drug repurposing. Front. Pharmacol. **13** (2022)
27. Saitoh, T., et al.: Hodgkin lymphoma presenting with various immunologic abnormalities, including autoimmune hepatitis, hashimoto's thyroiditis, autoimmune hemolytic anemia, and immune thrombocytopenia. Clin. Lymphoma Myeloma **8**(1), 62–64 (2008)
28. StellarGraph: Heterogeneous graphsage (hinsage) (2020). https://stellargraph.readthedocs.io/en/stable/hinsage.html
29. Sunghwan, K., et al.: PubChem in 2021: new data content and improved web interfaces. Nucleic Acids Res. (2021)
30. Wang, B., et al.: Similarity network fusion for aggregating data types on a genomic scale. Nat. Methods **11**(3), 333–337 (2014)
31. Wang, Z., Zhou, M., Arnold, C.: Toward heterogeneous information fusion: bipartite graph convolutional networks for in silico drug repurposing. Bioinformatics **36**(Supplement_1), i525–i533 (2020)
32. Weininger, D.: Smiles, a chemical language and information system. 1. Introduction to methodology and encoding rules. J. Chem. Inf. Comput. Sci. **28**(1), 31–36 (1988)
33. Yu, Z., Huang, F., Zhao, X., Xiao, W., Zhang, W.: Predicting drug-disease associations through layer attention graph convolutional network. Briefings Bioinform. **22**(4), bbaa243 (2021)
34. Zeng, X., Zhu, S., Liu, X., Zhou, Y., Nussinov, R., Cheng, F.: deepDR: a network-based deep learning approach to in silico drug repositioning. Bioinformatics **35**(24), 5191–5198 (2019)
35. Zhao, B.W., Hu, L., You, Z.H., Wang, L., Su, X.R.: HINGRL: predicting drug-disease associations with graph representation learning on heterogeneous information networks. Briefings Bioinform. **23**(1), bbab515 (2021)
36. Zhao, B.W., You, Z.H., Wong, L., Zhang, P., Li, H.Y., Wang, L.: MGRL: predicting drug-disease associations based on multi-graph representation learning. Front. Genet. **12**, 657182 (2021)

# Multiplex Collaboration Network of the Faculty of Computer Science and Engineering in Skopje

Ilinka Ivanoska$^{(\boxtimes)}$ ID, Kire Trivodaliev ID, and Bojan Ilijoski ID

Faculty of Computer Science and Engineering, Ss Cyril and Methodiuos University, Skopje, North Macedonia
{ilinka.ivanoska,kire.trivodaliev,bojan.ilijoski}@finki.ukim.mk

**Abstract.** Multiplex collaboration networks facilitate intricate connections among individuals, enabling multidimensional collaborations across various domains and fostering synergistic knowledge exchange. This study focuses on the construction and basic analysis of a multiplex collaboration network among employees at the Faculty of Computer Science and Engineering (FCSE), Ss. Cyril and Methodius University in Skopje. The multiplex network is built with three layers based on: scientific collaborations resulting from joint project participations by FCSE employees, joint employees participations in the FCSE graduation thesis committees, and scientific FCSE employees collaborations defined by co-authorships in Google Scholar papers.

The network's structure plays a vital role in determining the information accessibility and cooperative opportunities for individuals within FCSE institution. The aim here is to investigate the FCSE multiplex collaboration network's internal structure for discovering its latent knowledge and understand its implications. We perform identification of key individuals within the network, by computing various centrality and hubs detection network metrics. Additionally, we employ a community detection algorithm to reveal the underlying modular structure of the network.

By comprehensively analyzing the acquired multiplex collaboration network model, we contribute to a better understanding of the collaboration patterns among FCSE employees. The findings can potentially inform decision-making processes and foster strategic planning aimed at enhancing collaboration and knowledge sharing within the institution.

**Keywords:** multiplex networks · collaboration graph · centrality analysis · community structure · scientific project collaborations

## 1 Introduction

Collaboration plays a vital role in academic science, spanning across various fields and contributing to the expansion of research and publication endeavors

Supported by Faculty of Computer Science and Engineering, Skopje, N. Macedonia.

M. Mihova and M. Jovanov (Eds.): ICT Innovations 2023, CCIS 1991, pp. 206–221, 2024.
https://doi.org/10.1007/978-3-031-54321-0_14

[2,20,21]. It can be viewed as a process that facilitates the transfer of knowledge within scientific communities, providing individual scientists with opportunities to augment their own understanding. As collaboration networks grow, scientists gain access to information both directly, through their collaborators, and indirectly, through their collaborators' collaborators. The structure of these expansive networks can have subtle effects on the work of individual scientists, influencing their efforts in ways that may not be immediately apparent. Moreover, the overall structure of the collaboration network as a whole can impact scientific productivity. Certain network structures foster diverse and innovative work, while others may inadvertently lead to isolation and hinder the retention of creativity. Collaboration networks enable individual scientists to share knowledge and expertise, resulting in more efficient and impactful research outcomes [21,26]. By fostering professional relationships among researchers, collaboration networks facilitate potential future collaborations and interdisciplinary partnerships. Consequently, collaboration networks serve as a critical component of modern scientific research, playing a pivotal role in the advancement of scientific knowledge.

The study of diverse networks such as computer and information networks, social networks, and biological networks often reveals various network characteristics, including small-world phenomena, heavy-tailed distributions, and community structure [23,28,29]. Community structure refers to the presence of densely connected node clusters within a network, which are more interconnected internally than with other clusters. The presense of community structure and identification of communities within a network not only sheds light on its structure and functionality, but also promotes enhanced communication and collaboration within these communities [1,25,30]. This aspect holds particular importance in collaboration networks. One related problem is community collaborations search topic, which aims to identify communities associated with specific topics [15,31].

Also, numerous graph databases also include attributes associated with individual nodes, as well as different types of interaction. Simple approaches are unable to effectively incorporate this valuable multiple data into the networks analysis process. Multiplex networks [5,8,14,18], also known as interconnected networks, have gained significant attention in recent years due to their ability to capture and analyze complex systems with multiple types of interactions. Unlike traditional networks that represent relationships between entities using a single set of connections, multiplex networks consist of multiple layers, each representing a distinct type of relationship or interaction. By analyzing different layers of a multiplex network, researchers can uncover hidden patterns, identify influential nodes, and unravel the dynamics and cascading effects across multiple interconnected domains.

Multiplex collaboration networks [3] represent a specific type of multiplex network that focuses on capturing and analyzing collaborative relationships among entities in various areas. These networks capture the complexity and richness of collaborative interactions by considering multiple layers that represent different dimensions of collaboration, such as co-authorship, co-funding, co-patenting, or co-working. This framework allows researchers to explore the emergence of interdisciplinary collaborations, the impact of collaboration on knowledge transfer and innovation, and the identification of key actors and communities driving successful collaborations. Moreover, multiplex collaboration networks provide valuable insights into the design of effective strategies for fostering collaboration, allocating resources, and promoting cross-disciplinary interactions.

This paper focuses on building and analyzing the structure of the multiplex collaboration network among employees at the Faculty of Computer Science and Engineering (FCSE), Ss. Cyril and Methodius University in Skopje. The network is constructed based on three types of collaborations: 1) scientific projects collaborations which utilizes data from the portal http://projects.finki.ukim.mk, 2) graduation theses collaborations of data from the portal http://diplomski.finki.ukim.mk and 3) co-authorship scientific papers collaborations extracted from the Google Scholar database. The aim is to investigate the collaboration patterns among researchers within the FCSE community. The analysis includes examining the "affiliation" network, where researchers (FCSE employees) are associated with collaborative projects, graduation theses and co-authorship papers. Several metrics analysis (including centralities, hubs detection) is conducted to identify influential individuals within the network, and community structures are obtained to reveal cohesive groups of researchers.

The remaining sections of the paper are organized as follows. Section 2 provides a comprehensive explanation of the datasets utilized in this study, along with an overview of the system architecture developed for the analysis. Additionally, descriptions of the analysis metrics and community detection algorithms employed are presented in this section. The experimental setup, including the conducted experiments, results, and subsequent discussion, is outlined in Sect. 3. Finally, Sect. 4 summarizes the findings and offers concluding remarks.

## 2    Materials and Methods

In this section, we provide an overview of the datasets utilized to construct the multiplex collaboration network, as well as the system architecture and analysis algorithms employed.

### 2.1    Dataset

The datasets used in this study are sourced from: 1) the portal http://projects.finki.ukim.mk, which serves as the data source for building the FCSE projects collaboration network; 2) the portal http://diplomski.finki.ukim.mk as data source for building the FCSE graduation theses collaboration network; 3) the Google Scholar database as data source for building the co-authorship scientific papers network. These three sources provide valuable information on the collaborations among researchers within the FCSE community.

The portal http://projects.finki.ukim.mk, developed specifically for FCSE employees, serves as a repository for information regarding FCSE's national and international projects. The projects collaboration network is built based on the participation of individuals in the same projects, utilizing data extracted from this portal. Every representative FCSE project stores data for its name, description, date entered, project's principal investigator (PI), project members, project members' papers, goals and project output. http://diplomski.finki.ukim.mk is a portal developed at FCSE intended for its employees to store information about FCSE's students graduation theses. For each student graduation theses information about the mentor and two jury members which are FCSE employees, title and description is stored. The graduation theses collaboration network is built based on co-mentorship and collaboration on FCSE graduation theses derived from the second source, whereas, the last co-authorship scientific papers collaboration network is built based on papers co-authorship from the third source (the Google Scholar database).

## 2.2   System Architecture

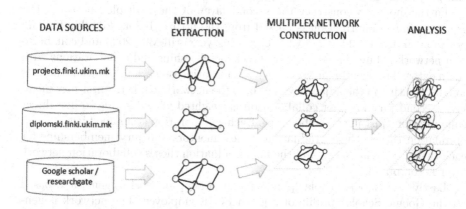

**Fig. 1.** FCSE system architecture for projects collaboration network analysis.

The system architecture devised for the FCSE collaboration network analysis is illustrated in Fig. 1. It comprises several modules: the data sources, the three networks layers extraction module, the multiplex network construction module and the networks analysis module. The data sources module retrieves relevant data from the aforementioned two portals and Google Scholar database, capturing the necessary details for constructing the collaboration networks. The three networks extraction module processes the retrieved data and generates comprehensive graph representations that encapsulate the relationships and connections among FCSE projects, graduation theses, papers and researchers. These graphs serve as foundation for subsequent analyses. The multiplex network module combines the three individual networks as layers and constructs the final network, whereas, the network analysis module focuses on conducting the analysis. By employing this system architecture, we can effectively extract, analyze, and interpret the collaboration patterns within the FCSE projects collaboration network.

## 2.3  Building the Multiplex Collaboration Network

To generate the first layer of the multiplex network i.e. the projects collaboration network, the adjacency matrix is constructed using direct collaborations and joint collaborations within FCSE projects. The network captures the relationships between individuals who have collaborated directly or indirectly on projects. The projects collaboration network is established by parsing the project information from the portal http://projects.finki.ukim.mk from 2011 (the FCSE beginning year) till May 2023. Memberships in projects and collaborations with the project's principal investigator (PI) are taken into account. Collaborations between the PI and project members are assigned a weight of 1 in the adjacency matrix, indicating direct collaboration. Collaborations between project members themselves are assigned a weight of 0.5, representing non-direct collaboration. The contributions from each collaboration are summed across all projects, PIs, and project members to obtain the final projects collaboration network.

Furthermore, to construct the second layer of the multiplex network, the graduate theses information is parsed from the http://diplomski.finki.ukim.mk portal again from 2011 (the FCSE beginning year) till May 2023 and collaboration network is built based on jury theses memberships and collaborations with the mentor. The mentor-jury member collaboration on each graduation thesis is weighted with 1 in the adjacency matrix of the network, while the first jury member - second jury member collaboration is weighted with 0.5, due to non-direct collaboration (just jury membership). Each collaboration is summed based on all graduation theses, these graduation theses' mentors and jury memberships for each two FCSE employees, and the final graduation theses collaboration network layer is computed.

Lastly, the Google Scholar papers co-authorship network is generated based on the Google Scholar profile of a given FCSE employee. The network is generated with its adjacency matrix according to direct collaboration and co-authorship on Google Scholar papers.

The resulting graph of the FCSE projects collaboration network comprises a total of 264 nodes, representing FCSE employees and their collaborators (including FCSE external collaborators, not FCSE employees). Among these nodes, 67 correspond to principal investigators (PIs) involved in the projects. Next, the FCSE resulting graduation theses collaboration graph has 98 nodes (FCSE employees and their external collaborators), out of which 63 correspond to graduation theses' mentors, while those and the remaining ones correspond to graduation theses jury members. Both constructed graphs exhibit a dense structure, indicating a significant level of collaboration among the network participants. Furthermore, the collaboration networks form a single connected component, indicating that all nodes in the graph are interconnected, allowing for seamless communication and collaboration across the network (see Fig. 2a) and b), as well as Sect. 4). Finally, the resulting Google scholar co-authorship papers graph is

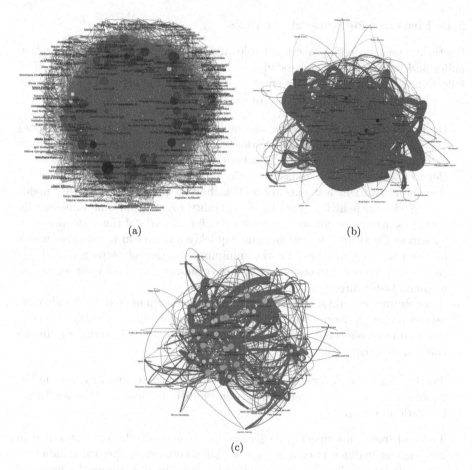

**Fig. 2.** a) (top left) FCSE projects, b) (top right) graduation theses and c) (bottom) google scholar co-authorship collaboration network.

consisted of 87 nodes, out of which 86 correspond to FCSE employees and 1 to other FCSE external collaborators. The last node is summed in node 'other' for simplicity, due to the large number of external collaborators (see Fig. 2c), as well as for enhanced networks visualization experience see Sect. 4). The multiplex network of these three resulting graphs is constructed with the assumption the different collaborations types are weighted with the same significance in the layers, not connecting the individual layers with weights, but just matching the same individuals in the three individual layers. For simplicity of the multiplex network analysis (explained in Sect. 3), the external FCSE collaborators of the first two layers, as well as administrative FCSE staff, are also filtered out as node 'other' in the same manner as it was done in the third layer graph construction. Finally, for the multiplex network we define the aggregated topological adjacency matrix which characterizes its connectivity and combines the adjacency matrices of the individual layers as in [4].

## 2.4    Measures for Network Analysis

Centrality measures play a crucial role in identifying the most collaborative nodes and influential researchers within the collaboration network [10,19]. The following types of centrality are employed to analyze the network (for all three individual layers, as well as the multiplex network):

- Degree Centrality: It quantifies the number of collaborations (connections) associated with each node [24]. Nodes with higher degree centrality are more extensively connected within the network.
- Betweenness Centrality: It assesses the importance of a node based on the frequency with which it appears on the shortest paths between other nodes [12]. Nodes with high betweenness centrality exert significant influence by acting as bridges or connectors between different parts of the network.
- Closeness Centrality: It evaluates how quickly a node can reach other nodes in the network, measured by the minimum number of steps required [6]. Nodes with higher closeness centrality have shorter distances to other nodes, enabling faster direct collaboration.
- Eigenvector Centrality: It measures a node's connections to other highly connected nodes [9]. Nodes with high eigenvector centrality have influence over multiple well-connected nodes, indicating a hidden level of control within the collaboration network.

Furthermore, the following metrics were examined to quantify the nodes' distribution among layers and to determine hub individuals for the multiplex collaboration network:

- The multiplex participation coefficient [4]: It assesses the level of a node's involvement in different communities within a network. Higher value indicates a more balanced distribution of the nodes participation across the layers of the multiplex. Truly multiplex are nodes for which the value is greater than 2/3, mixed the nodes for which the value is between 1/3 and 2/3, whereas, focused are those nodes with value between 0 and 1/3.
- Z-score of the overlapping degree [4]: It is used to distinguish hubs if it has a value greater than 2, from regular nodes, with a value less than 2.

By applying all these measures, we gain insights into the collaborative behavior and influence of individual researchers within the FCSE projects collaboration network.

## 2.5    Community Structure Detection

Community structure refers to the presence of clusters or groups of nodes in a network that exhibit higher levels of interconnectedness compared to the rest of the network. The identification of communities within a network can be challenging due to the unknown number of communities, variations in their size and density, and the absence of prior information about their existence. Despite these challenges, numerous approaches and methods have been developed to detect communities in networks: the Minimum-Cut Method [22], hierarchical clustering [16], Girvan-Newman Algorithm [13], Kernighan-Lin Algorithm [17], Walktrap Algorithm [27], modularity maximization [7,11] etc. The choice of method depends on the characteristics of the network and the specific goals of the analysis.

In this study, we employ the well-known Louvain community structure detection algorithm [7] to identify meaningful communities within the FCSE multiplex collaboration network (as well as the three individual layers) among employees/researchers. The Louvain algorithm is a widely used method for optimizing graph modularity [23]. By applying the Louvain algorithm, we aim to uncover cohesive groups of individuals who exhibit strong collaboration patterns within the FCSE projects collaboration network.

## 3   Experiments, Results and Discussion

### 3.1   Centrality Analysis

After constructing the FCSE collaboration network, we conducted a comprehensive centrality analysis of the multiplex network three layers. Figure 3 illustrates the results of this analysis, showcasing the four different centrality measures: degree (first graphs row), betweenness (second graphs row), closeness (third graphs row), and eigenvector centrality (fourth graphs row), for the projects (first graphs column), the graduation theses (second graphs column) and the co-authorship collaboration network (third graphs column), respectively. Each centrality graph represents the network's nodes and their connections, with the strength of the connections represented by the thickness of the corresponding edges. Additionally, the color and size of the nodes in Fig. 3 indicate their centrality metric values, with darker and larger nodes corresponding to higher centrality values. To enhance clarity, the visual representation of each graph has been thresholded, only displaying collaboration edges with weight values equal to or greater than a given value (see Sect. 4). This selection allows for a focused observation of the most significant collaborations within the FCSE projects collaboration network.

Table 1 presents a comprehensive overview of the top employees/researchers in the FCSE projects (first 5 people row), graduation thesis (second 5 people row) and co-authorship collaboration network (third 5 people row) based on the centrality analysis. The table displays the employees, ranked in descending order of their respective centrality metric. Similarly, the same table presents the top FCSE employees of the multiplex network centrality analysis (fourth 5 people row). Among the top researchers, examining the degree centrality metric, the top three researchers are K. Mishev, I. Chorbev, and B. Koteska, for the projects, graduation thesis and co-authorship networks layers, respectively. These individuals have the highest number of connections to other researchers in the network layer, which is reflected by their dark and large nodes in the first graphs row of Fig. 3. Similarly, S. Gievska, K. Trojachanec, and A. Mishev demonstrate

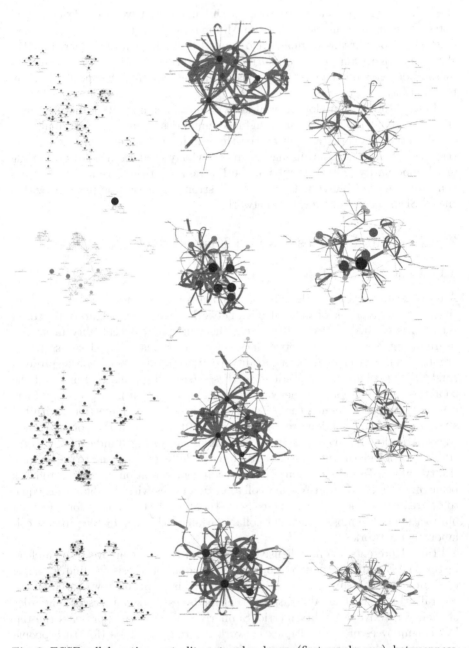

**Fig. 3.** FCSE collaboration centrality networks: degree (first graphs row), betweenness (second graphs row), closeness (third graphs row), and eigenvector centrality (fourth graphs row), for the projects (first graphs column), the graduation theses (second graphs column) and the co-authorship collaboration network (third graphs column), respectively.

**Table 1.** Summary table of top 5 employees/researchers in the FCSE collaboration network based on centrality and projects (first 5 people row), graduation thesis (second 5 people row), co-authorship (third 5 people row) and whole multiplex network (fourth 5 people row) analysis, sorted in descending order by each metric observed.

| Projects collaboration network | | | |
|---|---|---|---|
| Degree centrality | Betweennes centrality | Closeness centrality | Eigenvector centrality |
| K. Mishev | S. Gievska | K. Mishev | K. Mishev |
| S. Gievska | S. J. Sarknjac | S. Gievska | B. Jakimovski |
| B. Jakimovski | K. Zdravkova | B. Jakimovski | K. Kjiroski |
| B. Koteska | A. M. Bogdanova | B. Koteska | V. Bidikov |
| A. M. Bogdanova | K. Mishev | A. M. Bogdanova | G. Petkovski |
| Graduation theses collaboration network | | | |
| Degree centrality | Betweennes centrality | Closeness centrality | Eigenvector centrality |
| I. Chorbev | K. Trojachanec | I. Chorbev | I. Chorbev |
| M. Kostoska | P. Lameski | M. Kostoska | M. Kostoska |
| P. Lameski | E. Zdravevski | P. Lameski | P. Lameski |
| R. Stojanov | S. Filiposka | R. Stojanov | R. Stojanov |
| D. Trajanov | S. Kalajdziski | D. Trajanov | D. Trajanov |
| Co-authorship collaboration network | | | |
| Degree centrality | Betweennes centrality | Closeness centrality | Eigenvector centrality |
| B. Koteska | A. Mishev | B. Koteska | S. Filiposka |
| S. Filiposka | P. Lameski | S. Filiposka | B. Koteska |
| I. Chorbev | I. Dimitrovski | I. Chorbev | A. Kulakov |
| A. Mishev | V. D. Ristovska | A. Mishev | A. Mishev |
| A. Kulakov | M. Mirchev | A. Kulakov | A. M. Bogdanova |
| Multiplex collaboration network | | | |
| Degree centrality | Betweennes centrality | Closeness centrality | Eigenvector centrality |
| A. Kulakov | P. Lameski | A. Kulakov | V. Dimitrova |
| P. Lameski | A Kulakov | P. Lameski | A. Kulakov |
| V. Dimitrova | E. Zdraveski | V. Dimitrova | B. Koteska |
| E. Zdravevski | I. Chorbev | E. Zdraveski | P. Lameski |
| I. Chorbev | V. Dimitrova | I. Chorbev | S. Filiposka |

significant influence over others based on the betweenness centrality metric, as depicted in the second graphs row of Fig. 3. Furthermore, the same K. Mishev, I. Chorbev, and B. Koteska exhibit the fastest collaboration capabilities according to the closeness centrality analysis (third graphs row biggest and darkest nodes of Fig. 3). Lastly, K. Mishev, I. Chorbev, and S. Filiposka possess the highest hidden control within the individual three network layers, respectively, as indicated by the eigenvector centrality analysis (fourth graphs row of Fig. 3). In the multiplex network, A. Kulakov, P. Lameski and V. Dimitrova can be noticed as most influential from the centrality analysis.

Hence, the centrality analysis of the FCSE individual layers and the whole multiplex collaboration network reveals the presence of prominent individuals who consistently occupy influential positions across multiple centrality measures. Notably, names like K. Mishev, B. Jakimovski, S. Gievska, B. Koteska and A. M. Bogdanova in the projects network; I. Chorbev, M. Kostoska, P. Lameski, R. Stojanov and D. Trajanov in the graduation thesis network; and B. Koteska, A. Mishev, S. Filiposka, I. Chorbev and A. Kulakov in the co-authorship network emerge as key FCSE actors, respectively, demonstrating their significance as collaborators. These individuals consistently rank high across various centrality measures, some even in the multiplex network, indicating their substantial contributions and influence within the network. Their consistent prominence suggests that they play crucial roles in facilitating collaborations and knowledge exchange among researchers at FCSE. By being consistently positioned at the top across different centrality metrics, these "key" or "star" actors possess a strong presence and are likely to be instrumental in fostering effective collaborations and driving FCSE "affiliation" progress. Their contributions and leadership are essential in shaping the collaborative dynamics within the network and driving its overall success.

## 3.2   Multiplex Hubs Detection Analysis

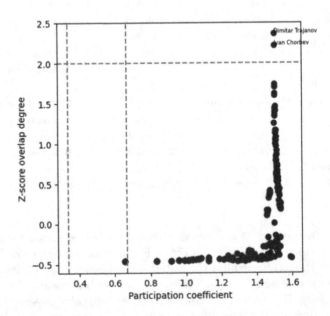

**Fig. 4.** FCSE multiplex collaboration network hubs detection.

The multiplex network analysis of the multiplex participation coefficient and the Z-score of the overlapping degree [4] show the distribution of nodes as regular or hubs, and as focused, mixed or truly multiplex. Figure 4 shows this analysis and detects the truly multiplex nodes (with a balanced distribution of participation across the layers of the multiplex), that are hubs: I. Chorbev and D. Trajanov. These two key FCSE employees have a high influence in the whole network, which corresponds in their power at the FCSE institution.

## 3.3    Community Detection Analysis

The calculated modularity of individual layers of the projects, graduation thesis and co-authorship collaboration networks, as well as the whole multiplex network are with a value of 0.175, 0.477 and 0.062, and 0.148, respectively. Modularity is a measure that evaluates the quality of the division of a network into communities. A higher modularity value indicates a stronger division of the network into separate communities, where nodes within the same community are densely connected to each other.

The results indicate a presence of a small non-trivial dense connected communities structure within the FCSE projects collaboration network and the whole multiplex. There is a significantly higher modularity in the graduation thesis collaboration network, which explains that FCSE employees form closed groups while being mentors and jury members on graduation thesis. On the other hand, a very low community structure in the co-authorship network is detected. The last low result can be explained due to vast amount of external collaborations on google scholar co-authorship papers, not taken into account. The existence of small dense connected communities implies that certain groups of individuals within the network are more interconnected with each other than with individuals from other communities. These communities likely represent cohesive research teams or subgroups within the FCSE that collaborate more frequently among themselves.

The FCSE collaboration network Louvain community detection algorithm analysis has revealed the presence of distinct communities. Figure 5 visualizes these communities for the projects (top left), graduation theses (top right) and co-authorship (bottom) network layers, with nodes of different colors representing individuals belonging to different communities. The visualization is generated by thresholding the graph in a similar manner as Fig. 3 for improved clarity. The size of the detected communities varies, ranging from small communities with only 2 nodes to larger communities with over 10 nodes. Additionally, the graph's node size in Fig. 5 corresponds to the level of influence that individuals hold within their respective communities. Nodes representing individuals with stronger influence appear larger in the visualization (see Sect. 4 for enhanced visualization experience). Interestingly, the leading researchers in all network layers are not concentrated within a single, exclusive community. Instead, they are distributed across different communities, indicating a diverse and interconnected network structure. This distribution of influential individuals throughout the network contributes to its stability, as the presence of key actors is not confined to a particular community. The detected communities align with the actual connections and groupings observed within the institution, reflecting

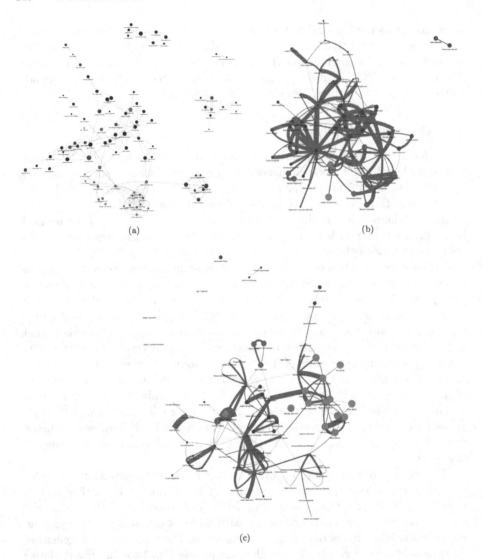

(a)

(b)

(c)

**Fig. 5.** a) (top left) FCSE projects, b) (top right) graduation theses and c) (bottom) google scholar co-authorship collaboration network communities discovered with Louvian community detection algorithm.

the real collaborative relationships among researchers. This finding underscores the relevance and accuracy of the community detection algorithm in identifying meaningful structures within the FCSE collaboration network.

# 4   Conclusion

The objective in this research has been twofold: 1) to build a FCSE multiplex collaboration network with three layers for projects, graduation thesis and google scholar co-authorship collaboration; and 2) to perform a basic analysis approach on the FCSE multiplex collaboration network based on centrality, hubs detection and community structure. Our in-depth centrality, hubs detection and community detection analysis indicate that there are no significant global changes occurring within the network. The prominent FCSE employees do not form an exclusive "closed" elite group, as multiple individuals (as I. Chorbev, D. Trajanov, P. Lameski, B. Jakimovski, A. Mishev, S. Gievska, B. Koteska, A. Kulakov, K. Mishev, S. Filipovska etc.) consistently rank highly across influential position measures. The highest and dense community structure is noticed in the graduation thesis collaborations, making those collaborations most closed ones.

Although the FCSE multiplex collaboration network is not large, the findings are valuable for illustrating potential future more novel approaches in measuring its structural multiplex network characteristics (including investigating small-world properties), in-depth. Our future research aims extend this research and explore the overlaps and causalities between the three layers of the multiplex collaboration network. We aim to investigate whether employees who serve together on a graduate thesis jury are more likely to collaborate and publish papers together in the future, or to propose a project together (and vice versa). Furthermore, we plan to expand the network semantically to explore whether a cluster of employees (professors) is elected as a jury graduation thesis member or a project member due to their shared field of work or other factors. We intend to leverage natural language processing techniques to analyze projects and graduation thesis titles and predict future collaborations and paper publications within the field of work of projects PIs, projects members, graduation theses jury members or mentors. Several questions arise here to be explored such as: what distinguishes individuals within different communities/clusters? Do they have numerous weak connections with other individuals, or do they maintain strong connections with specific characters while having weaker connections with others? How the individuals overlap in different collaborations types?

## Supplementary Material

The networks shown on Fig. 2, Fig. 3 and Fig. 5 are available on the repository https://github.com/bojanilijoski/fcse_colab_networks. The dataset used in this study is available upon request.

**Acknowledgements.** This work was partially financed by the Faculty of Computer Science and Engineering at the Ss. Cyril and Methodius University in Skopje.

# References

1. Airoldi, E.M., Blei, D., Fienberg, S., Xing, E.: Mixed membership stochastic block-models. Adv. Neural Inf. Process. Syst. **21**, 1–8 (2008)
2. Barabâsi, A.L., Jeong, H., Néda, Z., Ravasz, E., Schubert, A., Vicsek, T.: Evolution of the social network of scientific collaborations. Phys. A **311**(3–4), 590–614 (2002)
3. Battiston, F., Iacovacci, J., Nicosia, V., Bianconi, G., Latora, V.: Emergence of multiplex communities in collaboration networks. PLoS ONE **11**(1), e0147451 (2016)
4. Battiston, F., Nicosia, V., Latora, V.: Structural measures for multiplex networks. Phys. Rev. E **89**(3), 032804 (2014)
5. Battiston, F., Nicosia, V., Latora, V.: The new challenges of multiplex networks: measures and models. Eur. Phys. J. Spec. Topics **226**, 401–416 (2017)
6. Bavelas, A.: Communication patterns in task-oriented groups. J. Acoust. Soc. Am. **22**(6), 725–730 (1950)
7. Blondel, V.D., Guillaume, J.L., Lambiotte, R., Lefebvre, E.: Fast unfolding of communities in large networks. J. Stat. Mech. Theory Exp. **2008**(10), 1–12 (2008)
8. Boccaletti, S., et al.: The structure and dynamics of multilayer networks. Phys. Rep. **544**(1), 1–122 (2014)
9. Bonacich, P.: Factoring and weighting approaches to status scores and clique identification. J. Math. Sociol. **2**(1), 113–120 (1972)
10. Borgatti, S.P., Everett, M.G.: A graph-theoretic perspective on centrality. Social Netw. **28**(4), 466–484 (2006)
11. Fortunato, S.: Community detection in graphs. Phys. Rep. **486**, 75–174 (2010)
12. Freeman, L.C.: A set of measures of centrality based on betweenness. Sociometry **40**(1), 35–41 (1977)
13. Girvan, M., Newman, M.E.J.: Community structure in social and biological networks. Proc. Natl. Acad. Sci. **99**(12), 7821–7826 (2002)
14. Gomez, S., Diaz-Guilera, A., Gomez-Gardenes, J., Perez-Vicente, C.J., Moreno, Y., Arenas, A.: Diffusion dynamics on multiplex networks. Phys. Rev. Lett. **110**(2), 028701 (2013)
15. Himelboim, I., Smith, M.A., Rainie, L., Shneiderman, B., Espina, C.: Classifying twitter topic-networks using social network analysis. Social Media+ Soc. **3**(1), 2056305117691545 (2017)
16. Johnson, S.C.: Hierarchical clustering schemes. Psychometrika **32**(3), 241–254 (1967)
17. Kernighan, B.W., Lin, S.: An efficient heuristic procedure for partitioning graphs. Bell Syst. Tech. J. **49**(2), 291–307 (1970)
18. Kivelä, M., Arenas, A., Barthelemy, M., Gleeson, J.P., Moreno, Y., Porter, M.A.: Multilayer networks. J. Complex Netw. **2**(3), 203–271 (2014)
19. Landherr, A., Friedl, B., Heidemann, J.: A critical review of centrality measures in social networks. Bus. Inf. Syst. Eng. **2**(6), 371–385 (2010)
20. Newman, M.E.J.: Scientific collaboration networks. i. network construction and fundamental results. Phys. Rev. E **64**(1), 016131 (2001)
21. Newman, M.E.J.: The structure of scientific collaboration networks. Proc. Natl. Acad. Sci. **98**(2), 404–409 (2001)
22. Newman, M.E.J.: Detecting community structure in networks. Eur. Phys. J. B **38**, 321–330 (2004)
23. Newman, M.E.J.: Modularity and community structure in networks. Proc. Natl. Acad. Sci. **103**(23), 8577–8582 (2006)

24. Nieminen, J.: On the centrality in a graph. Scand. J. Psychol. **15**(1), 332–336 (1974)
25. Palla, G., Derényi, I., Farkas, I., Vicsek, T.: Uncovering the overlapping community structure of complex networks in nature and society. Nature **435**(7043), 814–818 (2005)
26. Paraskevopoulos, P., Boldrini, C., Passarella, A., Conti, M.: The academic wanderer: structure of collaboration network and relation with research performance. Ap. Net. Sci. **6**(1), 1–35 (2021)
27. Pons, P., Latapy, M.: Computing communities in large networks using random walks. J. Graph Algs. Appl. **10**(2), 191–218 (2006)
28. Porter, M.A., Onnela, J.P., Mucha, P.J.: Communities in networks. Not. Am. Math. Soc. **56**(9), 1082–1097 (2009)
29. Radicchi, F., Castellano, C., Cecconi, F., Loreto, V., Parisi, D.: Defining and identifying communities in networks. Proc. Natl. Acad. Sci. **101**(9), 2658–2663 (2004)
30. Traud, A.L., Kelsic, E.D., Mucha, P.J., Porter, M.A.: Comparing community structure to characteristics in online collegiate social networks. SIAM Rev. **53**(3), 526–543 (2011)
31. Zhao, Z., Feng, S., Wang, Q., Huang, J.Z., Williams, G.J., Fan, J.: Topic oriented community detection through social objects and link analysis in social networks. Knowl.-Based Syst. **26**, 164–173 (2012)

# Graph Neural Networks for Antisocial Behavior Detection on Twitter

Martina Toshevska$^{(\boxtimes)}$, Slobodan Kalajdziski, and Sonja Gievska

Faculty of Computer Science and Engineering, Ss Cyril and Methodiuos University,
Skopje, North Macedonia
{martina.toshevska,slobodan.kalajdziski,sonja.gievska}@finki.ukim.mk

**Abstract.** Social media resurgence of antisocial behavior has exerted a downward spiral on stereotypical beliefs, and hateful comments towards individuals and social groups, as well as false or distorted news. The advances in graph neural networks employed on massive quantities of graph-structured data raise high hopes for the future of mediating communication on social media platforms. An approach based on graph convolutional data was employed to better capture the dependencies between the heterogeneous types of data.

Utilizing past and present experiences on the topic, we proposed and evaluated a graph-based approach for antisocial behavior detection, with general applicability that is both language- and context-independent. In this research, we carried out an experimental validation of our graph-based approach on several PAN datasets provided as part of their shared tasks, that enable the discussion of the results obtained by the proposed solution.

**Keywords:** irony detection · hate speech detection · fake news detection · graph representation · heterogeneous graph · node classification · GraphSAGE · GAT · Graph Transformer

## 1 Introduction

With the rise of social media platforms, interpersonal communication has become easier and more frequent. However, antisocial behavior has also experienced an increase in various forms such as stereotypical or hateful comments toward individuals or social groups, false or distorted news, aggression, violence, etc. Although it could be beneficial for the author in terms of reaching more audiences or getting more views, likes, etc., it can be harmful to the target. Being able to detect online antisocial behavior could be a significant asset for social media platforms that enable them to perform actions to prevent it.

Graph Neural Networks (GNNs) are deep learning-based models that operate on graph structures. GNNs learn embedding representation for each node in the graph. Edge embeddings and graph embeddings can be created with the

---

Supported by Faculty of Computer Science and Engineering, Skopje, N. Macedonia.

M. Mihova and M. Jovanov (Eds.): ICT Innovations 2023, CCIS 1991, pp. 222–236, 2024.
https://doi.org/10.1007/978-3-031-54321-0_15

aggregation of node embeddings. GNNs perform two operations on the node embeddings obtained by the previous layer and the adjacency matrix of the graph [3,5,7,21]. The first operation is graph filtering which computes node embeddings, while the second is graph pooling which generates a smaller graph with fewer nodes and its corresponding new node embeddings. There is a variety of GNN models that implement various graph filtering functions.

In the past few years, GNNs have gained interest in the Natural Language Processing (NLP) field for text classification [10,22]. The traditional models based on recurrent neural networks (RNNs), convolutional neural networks (CNNs), and/or transformers capture contextual (local) information within a sentence. On the other hand, graph-based approaches capture global information about the vocabulary of a language [10]. Since the text data does not naturally have a graph structure, the crucial and most important part is to represent the text as a graph. Early approaches are focused on constructing text graphs composed of word nodes and documents nodes [22], while more recent approaches demonstrate that augmenting with additional information such as part of speech (POS) tags, named entities, and transformer-based word/sentence embeddings is beneficial.

In this paper, we evaluate the performance of several graph neural networks on the problem of detecting fake news and hate speech spreaders on Twitter[1] We define the problem as a node classification problem. We have created heterogeneous graphs using the datasets provided by a series of shared tasks on digital text forensics and stylometry (PAN) and we have trained several graph neural network models to classify user nodes. For comparison we have evaluated the proposed models on two additional tasks i.e. irony/stereotype spreaders on Twitter and sentiment classification on Yelp reviews. The rest of the paper is organized as follows. In Sect. 2, a brief introduction to GNN approaches to text classification problems is presented. The datasets are described in detail in Sect. 3. Section 4 presents the baseline models. The heterogeneous graph creation process is presented in Sect. 5, while Sect. 6 describes graph neural network models. The results are presented in Sect. 7. Section 8 concludes the paper.

## 2    Related Work

Graph-based approaches have been evaluated for many text classification tasks. TextGCN [22] operates on a heterogeneous graph created from text data representing words and documents as nodes, and relations between them as edges. Two-layer graph convolutional network (GCN) is applied on the heterogeneous text graph to allow indirect message passing between document nodes. TextGCN significantly outperforms baseline RNN-/CNN-based models on several benchmark datasets for sentiment classification, newsgroup classification, medical abstract classification, etc. The heterogeneous graph in our study was created following the TextGCN process of graph creation.

---

[1] The code for this research is available at: https://github.com/mtoshevska/ Antisocial-Behavior-on-Twitter.

VGCN-BERT [10] augments a BERT-based text classification model with graph embeddings to include global information about the vocabulary. A vocabulary graph has been constructed using normalized point-wise mutual information (NPMI). Vocabulary GCN (VGCN) has been applied to the vocabulary graph to create a graph embedding for the sentence. VGCN captures the part of the graph relevant to the input and then performs 2 layers of convolution, combining words from the input sentence with their related words in the vocabulary graph. To obtain the final class prediction, multiple layers of attention mechanism have been applied to the concatenated representation of the input text created with BERT and graph embeddings created with VGCN. VGCN-BERT has been evaluated on multiple text classification tasks including sentiment classification, hate speech detection, etc. In [8], a heterogeneous graph has been constructed following TextGCN [22], but a BERT/RoBERTa model has been used to obtain embeddings for the initial representation of the document nodes. The proposed model, BertGCN, has been optimized jointly with an auxiliary classifier that directly operates on BERT embeddings because it led to faster convergence and better performances. BertGCN parameters have been initialized with parameters of a pre-trained BERT model on the target dataset to speed up the training. Compared with the traditional BERT/RoBERTa models, the BertGCN yielded better performances. BertGCN has been evaluated on the same benchmark datasets as TextGCN. The performance gains obtained by BertGCN were higher for datasets containing longer sentences that enable capturing longer-term dependencies. Node representation in our study follows the BertGCN idea of document representation. We have utilized a BERT-based model to create an embedding for the initial representation of each tweet.

PAN[2] is a series of scientific events and shared tasks on digital text forensics and stylometry. There is a series of author profiling shared tasks that each year are focused on a different topic. In the past three years, they were focused on antisocial behavior detection on Twitter. The participants have used a wide variety of models starting from traditional machine learning models to Transformer-based architectures [12–14]. Most of the participants have used traditional machine learning approaches with various features such as n-grams, term frequency-inverse document frequency (TF-IDF), lexicons, word embeddings, sentence embeddings, etc. A few of the participants in 2020 [12] have created deep learning models such as multi-layer perceptron (MLP), CNNs, and RNNs. In the 2021 shared task [13], one of the participants built a BERT-based model with additional linear layers; and in 2022 [14], a graph convolutional neural network was first implemented by one of the participants. The best performing model for the task of fake news spreaders detection was a Logistic Regression model trained with n-gram features, as well as some statistic-based features from the tweets such as average length or lexical diversity [2]. The best performing model for the task of hate speech spreaders detection was a CNN model that used 100-dimensional word embedding vectors [18]. The best performing model for the task of detecting irony and stereotype spreaders was a CNN model with

---

[2] https://pan.webis.de/, last visited: 25.02.2023.

BERT-based tweet features [23]. In our experiments, we have used datasets provided by the PAN shared tasks. Since there was only one participant utilizing GNNs for the shared tasks, we aim to investigate in detail the performance of GNNs on these datasets.

# 3   Datasets

The datasets used for the experiments are provided by PAN for the Author Profiling shared tasks for the years 2020 (Profiling fake news spreaders on Twitter) and 2021 (Profiling hate speech spreaders on Twitter). The dataset for the 2022 shared task (Profiling irony and stereotype spreaders on Twitter - IROSTEREO) and a dataset for sentiment classification were also used to evaluate and compare the performances of the proposed models with more data.

## 3.1   Profiling Fake News Spreaders on Twitter

The training set provided in the Profiling Fake News Spreaders on Twitter[3] shared task is composed of 300 Twitter users with 100 tweets per user. Each user is labeled as either user posting tweets that contain fake news (1) or a user posting tweets that do not contain fake news (0). We have randomly chosen 80% of the users for training and 20% for validation in a way that the proportion of users in each class is retained. The testing set is composed of 200 Twitter users with 100 tweets per user.

## 3.2   Profiling Hate Speech Spreaders on Twitter

The training set provided in the Profiling Hate Speech Spreaders on Twitter[4] is composed of 200 Twitter users with 200 tweets per user. Each user is labeled as either user posting tweets that contain hate speech (1) or a user posting tweets that do not contain hate speech (0). Because the testing set was not available, we randomly split the users in the training set into subsets for training (80%), validation (10%), and testing (10%).

## 3.3   Profiling Irony and Stereotype Spreaders on Twitter - IROSTEREO

The training set provided in the Profiling Irony and Stereotype Spreaders on Twitter[5] shared task is composed of 420 Twitter users with 200 tweets per user. Each user is labeled as either user posting ironic tweets (I) or a user not posting

---

[3] https://pan.webis.de/clef20/pan20-web/author-profiling.html,     last     visited: 25.02.2023.

[4] https://pan.webis.de/clef21/pan21-web/author-profiling.html,     last     visited: 25.02.2023.

[5] https://pan.webis.de/clef22/pan22-web/author-profiling.html,     last     visited: 25.02.2023.

ironic tweets (NI). The testing set is composed of 180 Twitter users with 200 tweets per user. Although the testing set was available, the ground truth labels were not provided. We have created a training, validation, and testing subset by randomly choosing 80%, 10%, and 10% of the users, respectively.

In this shared task, another dataset for stereotype stance detection was provided. It contains the users that are labeled as users that are posting ironic tweets. Each user is labeled as either user posting ironic tweets with stereotypes in favor of the target (INFAVOR) or a user posting ironic tweets with stereotypes against the target (AGAINST). The training set is composed of 140 Twitter users, while the testing set is composed of 60 Twitter users. The number of tweets per user is 200. For this dataset, the testing set was available, but ground truth labels were not. We have created a training, validation, and testing subset by randomly choosing 80%, 10%, and 10% of the users, respectively.

### 3.4    Yelp Open Dataset

Yelp[6] dataset is a collection of 8.6 million business reviews that are rated with a 5-star rating system. We have created labels for the reviews according to the rating as follows. If the rating is less than or equal to 3 the review is labeled as negative, and as positive if the rating is greater than 3. The dataset was filtered in a way that the number of reviews per user is similar to the number of tweets per user in the previous datasets and the review length is similar to the tweet length. It has been filtered first by the number of reviews per user and then by the length of the reviews. We kept only the reviews written by users with a number of written reviews in the range from 50 to 200 with a length in the range from 15 to 60. Using the remaining reviews, we have created a training, validation, and testing subset by randomly choosing 80%, 10%, and 10% of the users, respectively.

## 4    Baseline Models

Following the success of the Transformer architectures for many natural language processing tasks and to compare the performance of the graph neural network models, we have trained three Transformer-based models: DistilBERT [16], RoBERTa [9], and DistilRoBERTa [15]. DistilBERT learns an approximate version of BERT using a knowledge distillation technique [1,6]. With only one-half of the layers of the original version of the BERT model, the number of parameters is reduced by 40%. DistilBERT is designed to be smaller and faster than BERT, while still retaining much of its accuracy. RoBERTa follows the original BERT architecture but has been trained with a different training procedure and on a larger corpus of text. It has been trained with dynamic masking where the masking pattern is generated every time a sequence is fed to the model, as opposed to static masking in the original BERT implementation where the same

---

[6] https://www.yelp.com/dataset, last visited: 25.02.2023.

training mask was used. RoBERTa has been trained without the next sentence prediction objective, with bigger batches over more data and longer sequences. DistilRoBERTa is a combination of the former two models. It learns an approximate version of the RoBERTa model following the same training procedure as in DistilBERT.

Since the goal is to classify users based on the tweets they have posted, we have concatenated all tweets of a particular user into one representation. We have used PyTorch implementation of these models available in the Huggingface Transformers library[7]. We initialized the weights with the pre-trained *distilbert-base-uncased*, *roberta-base*, and *distilroberta-base* weights for Distil-BERT, RoBERTa, and DistilRoBERTa models, respectively. All models have been trained with AdamW optimizer, binary cross-entropy loss, and batch size 16. For the other hyperparameters, we have performed a hyperparameter search among a set of possible values. The optimal hyperparameters for each model and each dataset are summarized in Table 1.

**Table 1.** Optimal hyperparameters for Transformer models.

| Fake News | | | |
|---|---|---|---|
| | **Learning Rate** | **Weight Decay** | **Epochs** |
| DistilBERT | 0.00001 | 0.005 | 100 |
| RoBERTa | 0.00001 | 0.00005 | 250 |
| DistilRoBERTa | 0.00001 | 0.005 | 250 |
| Hate Speech | | | |
| | **Learning Rate** | **Weight Decay** | **Epochs** |
| DistilBERT | 0.00001 | 0.0005 | 250 |
| RoBERTa | 0.00001 | 0.0005 | 250 |
| DistilRoBERTa | 0.00001 | 0.0005 | 250 |
| Irony Stereotype | | | |
| | **Learning Rate** | **Weight Decay** | **Epochs** |
| DistilBERT | 0.00001 | 0.005 | 500 |
| RoBERTa | 0.00001 | 0.0005 | 100 |
| DistilRoBERTa | 0.00001 | 0.005 | 100 |
| Stereotype Stance | | | |
| | **Learning Rate** | **Weight Decay** | **Epochs** |
| DistilBERT | 0.00001 | 0.005 | 500 |
| RoBERTa | 0.00001 | 0.005 | 100 |
| DistilRoBERTa | 0.0001 | 0.005 | 100 |
| Yelp | | | |
| | **Learning Rate** | **Weight Decay** | **Epochs** |
| DistilBERT | 0.00001 | 0.005 | 500 |
| RoBERTa | 0.00001 | 0.00005 | 250 |
| DistilRoBERTa | 0.00001 | 0.0005 | 250 |

---

[7] https://huggingface.co/docs/transformers/index, last visited: 25.02.2023.

## 5   Heterogeneous Graph Creation

We have created a heterogeneous graph dataset for classifying Twitter users, composed of three types of nodes: (1) user nodes, (2) tweet nodes, and (3) word nodes; and four types of edges: (1) user-tweet, (2) tweet-word, (3) word-word, and (4) tweet-tweet. The graph was created using all data in the subsets for training, validation, and testing. A simplified visualization of the graph is shown in Fig. 1.

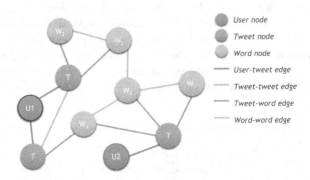

**Fig. 1.** Simplified visualization of the heterogeneous graph. $U1$, $U2$ - user nodes. $P$ - tweet nodes. $W_1$-$W_5$ - word nodes. The user $U1$ represents a user from the first class (e.g. posting ironic tweets), while the user $U2$ represents a user from the second class (e.g. not posting ironic tweets).

A vocabulary composed of the unique words in the dataset has been created. Special tokens representing user mentions, links, and hashtags have been added to the vocabulary. Rare words (words with less than 15 occurrences) have been removed and the remaining were used as word nodes.

Following the BertGCN [8] model, we utilize word and sentence embeddings to encode the nodes. Each word node is initialized with a word embedding of the corresponding word. We have used 200-dimensional GloVe [11] embedding vectors pre-trained on a Twitter dataset. The embeddings have been extracted using the Gensim library[8].

Each tweet is represented as a node initialized with a 768-dimensional sentence embedding obtained by a pre-trained DistilRoBERTa [15] model. The embeddings have been obtained using the Sentence-Transformers library[9]. User nodes have been initialized via the embedding representation of their tweets. Pre-trained DistilRoBERTa embeddings have been obtained for each of the 200 tweets per user. The embeddings have been averaged along the 0-axis thus ending with a 768-dimensional representation for each user.

---

[8] https://radimrehurek.com/gensim/, last visited: 25.02.2023.
[9] https://www.sbert.net/, last visited: 25.02.2023.

Word-word and tweet-word edges have been added following the graph creation process for the TextGCN model [22]. Edges between a pair of word nodes are added if the PMI is greater than 0. PMI value has been set as a weight for word-word edges. Edges between words and tweets are added with the TF-IDF of the word in the tweet as a weight for the edge. User-tweet edges have been added between each user and their 200 tweets. Tweet-tweet edges have been added following the CLHG [20] model. Each tweet is linked with the K most similar tweets according to cosine similarity (we set the value for K to 3). The cosine similarity was computed on the corresponding 768-dimensional DistilRoBERTa sentence embeddings.

The total number of nodes and edges for each dataset is summarized in Table 2.

**Table 2.** Number of nodes and edges in the created heterogeneous graphs for each of the four datasets.

|  | FN | HS | IS | SS | Y |
|---|---|---|---|---|---|
| User nodes | 500 | 200 | 420 | 140 | 883 |
| Tweet nodes | 50,000 | 40,000 | 84,000 | 28,000 | 68,172 |
| Word nodes | 3,506 | 2,713 | 8,580 | 3,394 | 5,557 |
| Total | 54,006 | 42,913 | 93,000 | 31,534 | 74,612 |
| User-tweet edges | 50,000 | 40,000 | 84,000 | 28,000 | 68,172 |
| Tweet-tweet edges | 150,000 | 120,000 | 252,000 | 84,000 | 204,516 |
| Tweet-word edges | 454,244 | 326,363 | 1,563,131 | 409,909 | 1,807,498 |
| Word-word edges | 278,668 | 187,540 | 1,020,308 | 263,862 | 592,033 |
| Total | 932,912 | 673,903 | 2,919,439 | 785,771 | 2,672,219 |

# 6    Graph Neural Network Models

In this research, three GNN architectures have been investigated for antisocial behavior detection: GraphSAGE [5], Graph Attention Network (GAT) [19], and Graph Transformer [4,17]. GraphSAGE is an inductive methodology for graph representation learning using sampling and aggregation of features from a node's local fixed-size neighborhood. Different tasks and problems are likely to leverage different aggregation functions (e.g., mean, LSTM pooling) and/or loss functions. GAT leverages masked self-attention layers in graph neural networks. The hidden representation of the nodes is computed with a self-attention mechanism that enables the nodes to attend to neighborhood features by specifying different weights for each neighbor node. Graph Transformer [17] is a generalization of the Transformer architectures for graph structures. The attention mechanism is represented as a function of the neighborhood connectivity for each node in the graph and the positional encoding is represented by the Laplacian eigenvectors. The normalization layer is replaced by a batch normalization layer. The architecture could be extended to edge feature representation. Unified Message Passing (UniMP) [4] jointly performs feature and label propagation by embedding the partially observed labels into the same space as node features. It is trained with a masked label prediction strategy inspired by BERT. We have used the modified Graph Transformer operator from the UniMP.

Our architecture is composed of a two-layer heterogeneous graph neural network followed by a ReLU activation that maps the nodes into a low-dimensional latent space. For the purpose of classifying nodes, a fully-connected layer has been added on top of the GNN model, which infers the class for the user nodes. The architecture is the same for all three models and is displayed in Fig. 2.

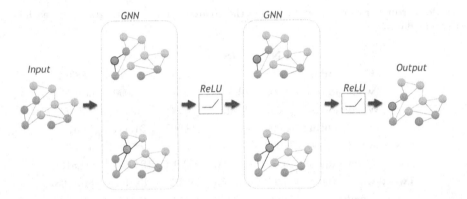

**Fig. 2.** Architecture of a heterogeneous GNN model.

We have used PyTorch implementation of these models available in the PyTorch Geometric library[10]. The models have been created with the implementation for homogeneous graphs, and then are transformed into models suitable for heterogeneous graphs. All models have been trained with AdamW optimizer and binary cross-entropy loss. For the other hyperparameters, we have performed a hyperparameter search among a set of possible values. The optimal hyperparameters for each model and each dataset are summarized in Table 3.

## 7    Results

We have performed several experiments with baseline Transformer models and GNN models. For each dataset, we have trained six models with the corresponding optimal hyperparameters shown in Table 1 and Table 3. Each of the models has been trained on Quadro RTX 8000 48GB GPU.

---

[10] https://pytorch-geometric.readthedocs.io/en/latest/, last visited: 25.02.2023.

**Table 3.** Optimal hyperparameters for GNN models.

| Fake News | | | |
|---|---|---|---|
| | Learning Rate | Weight Decay | Epochs |
| GraphSAGE | 0.01 | 0.00005 | 250 |
| GAT | 0.001 | 0.0005 | 250 |
| GraphTransformer | 0.01 | 0.00005 | 500 |
| **Hate Speech** | | | |
| | Learning Rate | Weight Decay | Epochs |
| GraphSAGE | 0.01 | 0.0005 | 50 |
| GAT | 0.01 | 0.005 | 250 |
| GraphTransformer | 0.001 | 0.00005 | 50 |
| **Irony Stereotype** | | | |
| | Learning Rate | Weight Decay | Epochs |
| GraphSAGE | 0.01 | 0.05 | 50 |
| GAT | 0.0001 | 0.00005 | 250 |
| GraphTransformer | 0.001 | 0.05 | 250 |
| **Stereotype Stance** | | | |
| | Learning Rate | Weight Decay | Epochs |
| GraphSAGE | 0.01 | 0.05 | 50 |
| GAT | 0.01 | 0.005 | 50 |
| GraphTransformer | 0.01 | 0.05 | 250 |
| **Yelp** | | | |
| | Learning Rate | Weight Decay | Epochs |
| GraphSAGE | 0.01 | 0.0005 | 100 |
| GAT | 0.0001 | 0.0005 | 500 |
| GraphTransformer | 0.0001 | 0.05 | 100 |

## 7.1   Comparison with Baseline Models

To evaluate the models, accuracy has been calculated for the samples in the corresponding test sets. The results are shown in Table 4. Evaluation results of the three best performing models in the shared tasks are also included. For the fake news dataset, the test set provided by PAN has been used. For the other datasets, 10% of the training set has been utilized for testing. The subtask of stereotype stance detection has been evaluated with the F1 measure. The results of the best performing models on this dataset are not shown since our models were not evaluated with the F1 measure.

The results show that for most of the cases, GNN models, in general, perform worse than the baseline Transformer models. Since deep neural networks require huge amounts of data for training and given that these datasets are relatively small, we could hypothesize that the worse performance is due to the small amount of data. On the other hand, Transformer-based models are pre-trained on large datasets which gives them a significant advantage over the other models.

For the hate speech dataset, both GraphSAGE and GraphTransformer models have the same accuracy as DistilBERT which is the second best model for the dataset. For the stereotype stance dataset, the GraphTransformer model has the same accuracy as RoBERTa which is the third best model for the dataset. The difference from the best performing model is 0.1 for the hate speech dataset and 0.14 for the stereotype stance dataset. These results demonstrate the capability of GNN models to successfully learn from graphs created from text data.

To compare with the best performing models in the PAN shared tasks, all three GNN models outperform the three best performing models for the hate speech dataset. However, for the fake news and irony stereotype datasets, the performance is inferior. The hate speech dataset is the smallest one among the three. Taking into account the fact that the models were not pre-trained, we could hypothesize that learning from a smaller graph is easier when the models are not pre-trained. Transformer-based models outperform the baseline models for the fake news and hate speech datasets. The DistilRoBERTa model has the best performance on the fake news dataset, while the RoBERTa model is the best performing model on the hate speech dataset. Nevertheless, we should point out that for all the datasets, except the fake news dataset, the evaluation was not done using the same test set, and therefore we could not know precisely how they would perform if the original test set was used.

**Table 4.** Evaluation results and comparison with baseline models. The metric shown is accuracy. For the Stereotype Stance dataset participants were ranked according to the F1 measure and the results are not shown here.

| | Fake News | Hate Speech | Irony Stereotype | Stereotype Stance | Yelp |
|---|---|---|---|---|---|
| DistilBERT | 0.72 | 0.80 | 0.83 | 0.93 | 0.74 |
| RoBERTa | 0.72 | 0.90 | 0.90 | 0.79 | 0.78 |
| DistilRoBERTa | 0.80 | 0.70 | 0.83 | 0.86 | 0.73 |
| GraphSAGE | 0.54 | 0.80 | 0.60 | 0.71 | 0.63 |
| GAT | 0.56 | 0.75 | 0.74 | 0.71 | 0.62 |
| GraphTransformer | 0.55 | 0.80 | 0.67 | 0.79 | 0.64 |
| #1 | 0.75 | 0.74 | 0.99 | / | / |
| #2 | 0.75 | 0.73 | 0.98 | / | / |
| #3 | 0.74 | 0.72 | 0.97 | / | / |

## 7.2  Ablation Study

To analyze the effectiveness of each component in the graph, ablation studies have been performed. Four variants of the heterogeneous graph have been examined:

- *all* - all components are included.
- *no-word-word* - edges between word nodes are excluded from the graph.
- *no-word* - word nodes and edges that they are part of are excluded from the graph.
- *no-doc-doc* - edges between tweet nodes are excluded from the graph.

A separate model has been trained using the optimal hyperparameters for each variant and accuracy on the test set was calculated. The results are summarized in Table 5.

The results show that the best performance is achieved when all the components in the graph are included. One exception is the GraphSAGE model on the irony stereotype dataset for which the best performance is achieved by the *no-word* variant suggesting that removing word nodes and edges that they are part of leads to better performance than including all the components in the graph. This dataset is significantly bigger than the others and we can conclude that user and tweet nodes, as well as edges between them, are sufficient for the GraphSAGE model to successfully learn to classify the users. For the stereotype stance dataset, the GraphSAGE model achieved the same performances for all variants. The worst performance for all datasets is achieved with the *no-doc-doc* variant indicating that removing the edges between tweet pairs reduces the performance. Edges between tweet pairs add shortcuts in the processing that could lead to faster convergence of the models and we could expect worse performances with their removal. An exception is the GraphTransformer model on the Yelp dataset for which the *no-doc-doc* variant achieves the second best result. The structure of the reviews in the Yelp dataset differs from the tweets in the other Twitter datasets.

For the GraphSAGE model, the *no-word* variant is better than the *no-word-word* variant for all Twitter datasets suggesting that removing any component that is related to words is better than removing only edges between word pairs. For the Yelp dataset, the *no-word-word* variant is better. We could hypothesize that removing only edges between word pairs leads to better results for the Yelp reviews rather then removing any word related component. GraphTransformer follows the same pattern except for the stereotype stance dataset for which the *no-word-word* variant achieved better results. GAT model has better results with the *no-word-word* variant for the fake news and irony stereotype datasets, while the *no-word* variant is better for the hate speech and stereotype stance datasets. The latter are smaller datasets and we could hypothesize that the GAT model could learn to better classify user nodes without any component related to words for smaller datasets.

**Table 5.** Ablation results for the GNN models. The shown values represent the accuracy metric on the test set.

| Fake News | | | | |
|---|---|---|---|---|
| | **all** | **no-word-word** | **no-word** | **no-doc-doc** |
| GraphSAGE | **0.54** | 0.52 | 0.54 | 0.51 |
| GAT | **0.56** | 0.51 | 0.50 | 0.47 |
| GraphTransformer | **0.55** | 0.46 | 0.54 | 0.44 |

| Hate Speech | | | | |
|---|---|---|---|---|
| | **all** | **no-word-word** | **no-word** | **no-doc-doc** |
| GraphSAGE | **0.80** | 0.65 | 0.75 | 0.65 |
| GAT | **0.75** | 0.60 | 0.70 | 0.65 |
| GraphTransformer | **0.80** | 0.70 | 0.70 | 0.70 |

| Irony Stereotype | | | | |
|---|---|---|---|---|
| | **all** | **no-word-word** | **no-word** | **no-doc-doc** |
| GraphSAGE | 0.60 | 0.57 | **0.62** | 0.57 |
| GAT | **0.74** | 0.55 | 0.50 | 0.50 |
| GraphTransformer | **0.67** | 0.60 | 0.64 | 0.57 |

| Stereotype Stance | | | | |
|---|---|---|---|---|
| | **all** | **no-word-word** | **no-word** | **no-doc-doc** |
| GraphSAGE | **0.71** | 0.71 | 0.71 | 0.71 |
| GAT | **0.71** | 0.50 | 0.71 | 0.71 |
| GraphTransformer | **0.79** | 0.79 | 0.57 | 0.57 |

| Yelp | | | | |
|---|---|---|---|---|
| | **all** | **no-word-word** | **no-word** | **no-doc-doc** |
| GraphSAGE | **0.63** | 0.56 | 0.55 | 0.42 |
| GAT | **0.62** | 0.55 | 0.53 | 0.47 |
| GraphTransformer | **0.64** | 0.49 | 0.49 | 0.56 |

## 8   Conclusion

This paper explored the performances of graph neural networks for the task of antisocial behavior detection on Twitter. Three GNN architectures (Graph-SAGE, GAT, and Graph Transformer) were evaluated against four datasets composed of Twitter users and tweets that they have posted that were provided by PAN shared tasks, and one dataset composed of Yelp users and reviews that they have written that was extracted from the Yelp Open Dataset. A heterogeneous graph dataset has been created with user, tweet/review, and word nodes, as well as five types of edges between them.

An ablation study was performed to investigate which components of the heterogeneous graph have contributed the most. The results showed that the best performances are achieved when all the graph components are included, while the worst performances were obtained when the edges between tweet pairs were excluded from the graph.

Transformer-based models were also trained on the same datasets as baseline models for comparison. When compared against the baseline models, the GNN models showed inferior performance for most of the experiments. For the experiments, pre-trained Transformer-based models (DistilBERT, RoBERTa, and DistilRoBERTa) have been used. The models are pre-trained on large datasets which gives them a significant advantage. This hypothesis leads to a possible future direction which is to first pre-train GNNs on a larger dataset, and then train on the specific datasets that were used in this research.

For two of the datasets employed in this study, GNN models showed comparable performances with second best and third best Transformer-based models. These findings indicate the capability of GNN models to learn from the types of data derived from social networks that were utilized in this research. We anticipate that GNN models could be successfully applied to other text classification or even wider natural language processing or generation tasks.

**Acknowledgements.** This work was partially financed by the Faculty of Computer Science and Engineering at the Ss. Cyril and Methodius University in Skopje.

# References

1. Bucilǎ, C., Caruana, R., Niculescu-Mizil, A.: Model compression. In: Proceedings of the 12th ACM SIGKDD International Conference on Knowledge Discovery and Data Mining, pp. 535–541 (2006)
2. Buda, J., Bolonyai, F.: An Ensemble Model Using N-grams and Statistical Featuresto Identify Fake News Spreaders on Twitter-Notebook for PAN at CLEF 2020 (2020)
3. Defferrard, M., Bresson, X., Vandergheynst, P.: Convolutional neural networks on graphs with fast localized spectral filtering. Adv. Neural Inf. Process. Syst. **29** (2016)
4. Dwivedi, V.P., Bresson, X.: A generalization of transformer networks to graphs. In: AAAI Workshop on Deep Learning on Graphs: Methods and Applications (2021)
5. Hamilton, W.L., Ying, R., Leskovec, J.: Inductive representation learning on large graphs. In: Proceedings of the 31st International Conference on Neural Information Processing Systems, pp. 1025–1035 (2017)
6. Hinton, G., Vinyals, O., Dean, J.: Distilling the knowledge in a neural network. arXiv preprint arXiv:1503.02531 (2015)
7. Kipf, T.N., Welling, M.: Semi-supervised classification with graph convolutional networks. In: 5th International Conference on Learning Representations, ICLR 2017, Toulon, France, 24–26 April 2017, Conference Track Proceedings (2017)
8. Lin, Y., et al.: Bertgcn: transductive text classification by combining GNN and Bert. In: Findings of the Association for Computational Linguistics: ACL-IJCNLP 2021, pp. 1456–1462 (2021)
9. Liu, Y., et al.: Roberta: a robustly optimized bert pretraining approach. arXiv preprint arXiv:1907.11692 (2019)

10. Lu, Z., Du, P., Nie, J.Y.: VGCN-BERT: augmenting BERT with graph embedding for text classification. In: Jose, J., et al. (eds.) Advances in Information Retrieval. ECIR 2020. LNCS, vol. 12035, pp. 369–382. Springer, Cham (2020). https://doi.org/10.1007/978-3-030-45439-5_25

11. Pennington, J., Socher, R., Manning, C.D.: Glove: global vectors for word representation. In: Proceedings of the 2014 Conference on Empirical Methods in Natural Language Processing (EMNLP), pp. 1532–1543 (2014)

12. Rangel, F., Giachanou, A., Ghanem, B.H.H., Rosso, P.: Overview of the 8th author profiling task at pan 2020: profiling fake news spreaders on twitter. In: CEUR Workshop Proceedings, vol. 2696, pp. 1–18. Sun SITE Central Europe (2020)

13. Rangel, F., Peña-Sarracén, G.L.D.L., Chulvi-Ferriols, M.A., Fersini, E., Rosso, P.: Profiling hate speech spreaders on twitter task at pan 2021. In: Proceedings of the Working Notes of CLEF 2021, Conference and Labs of the Evaluation Forum, Bucharest, Romania, 21st to 24th September 2021, pp. 1772–1789. CEUR (2021)

14. Reynier, O.B., Berta, C., Francisco, R., Paolo, R., Elisabetta, F.: Profiling irony and stereotype spreaders on twitter (IROSTEREO) at pan 2022. CEUR-WS. org (2022)

15. Sajjad, H., Dalvi, F., Durrani, N., Nakov, P.: On the effect of dropping layers of pre-trained transformer models. CoRR (2020)

16. Sanh, V., Debut, L., Chaumond, J., Wolf, T.: Distilbert, a distilled version of bert: smaller, faster, cheaper and lighter. arXiv preprint arXiv:1910.01108 (2019)

17. Shi, Y., Huang, Z., Feng, S., Zhong, H., Wang, W., Sun, Y.: Masked label prediction: unified message passing model for semi-supervised classification. CoRR (2020)

18. Siino, M., Di Nuovo, E., Tinnirello, I., La Cascia, M.: Detection of hate speech spreaders using convolutional neural networks–Notebook for PAN at CLEF 2021 (2021)

19. Veličković, P., Cucurull, G., Casanova, A., Romero, A., Liò, P., Bengio, Y.: Graph attention networks. In: International Conference on Learning Representations (2018). https://openreview.net/forum?id=rJXMpikCZ, accepted as poster

20. Wang, Z., Liu, X., Yang, P., Liu, S., Wang, Z.: Cross-lingual text classification with heterogeneous graph neural network. In: Proceedings of the 59th Annual Meeting of the Association for Computational Linguistics and the 11th International Joint Conference on Natural Language Processing (Volume 2: Short Papers), pp. 612–620 (2021)

21. Wu, L., et al.: Graph neural networks for natural language processing: a survey. CoRR (2021)

22. Yao, L., Mao, C., Luo, Y.: Graph convolutional networks for text classification. In: Proceedings of the AAAI Conference on Artificial Intelligence, vol. 33, pp. 7370–7377 (2019)

23. Yu, W., Boenninghoff, B., Kolossa, D.: BERT-based ironic authors profiling (2022)

# Theoretical Informatics

# On the Construction of Associative and Commutative Bilinear Multivariate Quadratic Quasigroups of Order $2^n$

Marija Mihova[(✉)]

Faculty of Computer Science and Engineering, Ss Cyril and Methodiuos University, Skopje, North Macedonia
marija.mihova@finki.ukim.mk

**Abstract.** Quasigroups have various applications in mathematics, computer science, and cryptography. In coding theory and cryptography they have been used in error-correcting codes, error-detection codes, to construct key exchange protocols and cryptographic primitives. There are also used in graph theory, experimental design and combinatorial designs. Quasigroups of order $2^n$ can be represented as vector valued Boolean functions from $\{0,1\}^n \times \{0,1\}^n$ to $\{0,1\}^n$. When the order of each coordinate functions is at most two, they are called Multivariate Quadratic Quasigroups (MQQ). In this paper we give a description of the functions representing Bilinear MQQ quasigroups, with a special focus on quasigroups that are commutative or associative.

**Keywords:** Quasigroups · Multivariate Quadratic Quasigroups · commutative · associative · Bilinear MQQ

## 1 Introduction

Quasigroups, also known as latin squares, are intriguing objects of study for researchers due to their unique algebraic properties, connections to graph theory, combinatorics and designs [1,2], applications in coding theory and cryptography [3–5], and potential applications in diverse fields. The exploration of quasigroups contributes to the broader understanding of algebraic structures and their applications in various disciplines. They provide researchers with a rich area of study to explore algebraic properties, operations, [6] and their implications [7,8].

MMQ quasigroups were introduced by Danilo [9] as block ciphers, but they quickly exhibited vulnerability to attacks [10,11]. This contributes to facilitating a more in-depth analysis of the structure of such quasigroups. This encourages a deeper analysis of such quasigroup structures [12–15]. This paper extends the results from the paper [14] and analyzes the form of matrices that represented MQQ that possess commutative or associative property, due to their applicability. Key exchange protocols typically require an operation with associativity in

Supported by Faculty of Computer Science and Engineering, Skopje, N. Macedonia.

M. Mihova and M. Jovanov (Eds.): ICT Innovations 2023, CCIS 1991, pp. 239–250, 2024.
https://doi.org/10.1007/978-3-031-54321-0_16

order to ensure that the exchanged keys are valid and consistent among participants. The associativity property ensures that the order in which the operations
are performed does not affect the final result, allowing participants to independently compute the same shared key. Commutativity is not a property of crucial
importance in applications in cryptography, but still has its place in certain
mathematical contexts.

The paper is organized as in six sections. In the second chapter of this work,
we present crucial results from our prior studies that form the foundation for
the subsequent progression of our research. The third chapter is consists of a
collection of novel results that serve as an extension of the findings presented in
the second chapter. The fourth and fifth chapters of this work are dedicated to
presenting a range of novel theorems that focus on describing the form of matrices
used to define associative and commutative quasigroups. The final chapter of
this study serves as a culmination of our research, providing a comprehensive
conclusion and outlining future directions for further investigation.

## 2     Matrix Formulations of BMQQ

In this chapter, we outline key findings from our earlier studies, which serve as a
basis for our subsequent analysis. All theorems from this chapter a given without
proof, since there are proven in [14].

**Definition 1.** *The quasigroup* $(\{0,1\}^n, *)$ *is called a Bilinear Multivariate
Quadratic Quasigroups (BMQQ) if the quasigroup operation can be represented
by a vector valued Boolean function* $f(\mathbf{x}, \mathbf{y}) = \mathbf{z}$, *where the k-th coordinate of* $\mathbf{z}$
*is defined as:*

$$z_k = c_k + \sum_{i=1}^{n} a_{k,i} x_i + \sum_{i=1}^{n} b_{k,i} y_i + \sum_{i=1}^{n} \sum_{j=1}^{n} d_{k,i,j} x_i y_j \tag{1}$$

*Terms* $c_k, a_{k,i}, b_{k,i}, d_{k,i,j} \in \{0,1\}, k, i, j = \overline{1,n}$ *are constants.*

*Using following notation:* $A = [a_{k,i}]$, $B = [b_{k,i}]$, $\mathbf{c} = [c_k]$ , $D'_i = \left[d'^{i}_{k,j}\right]$, *and*
$D''_j = \left[d''^{j}_{k,i}\right]$ *where* $d'^{i}_{k,j} = d''^{j}_{k,i} = d_{k,i,j}$, $\mathbf{x} * \mathbf{y}$ *can be represented as:*

$$\mathbf{x} * \mathbf{y} = \mathbf{c} + A\mathbf{x} + B\mathbf{y} + \sum_{i=1}^{n} x_i D'_i \mathbf{y} = \mathbf{c} + A\mathbf{x} + B\mathbf{y} + \sum_{j=1}^{n} y_j D''_j \mathbf{x}. \tag{2}$$

*When* $A = B = I$, *the quasigroup is called normalized BMQQ, (nBMQQ).
The matrix form of nBMQQ is:*

$$\mathbf{x} * \mathbf{y} = \mathbf{x} + \mathbf{y} + \sum_{i=1}^{n} x_i D'_i \mathbf{y} = \mathbf{x} + \mathbf{y} + \sum_{j=1}^{n} y_j D''_j \mathbf{x}$$

It is evident that the $j$-th column of the matrix $D_i'$ is equal to the $i$-th column of the matrix $D_j''$, so each nBMQQ is fully specified by one of the vectors of the binary $n \times n$ matrices $\mathbf{D}' = (D_1', \ldots, D_n')$ and $\mathbf{D}'' = (D_1'', \ldots, D_n'')$. In general case, it is necessary to specify matrices $A$ and $B$. The first vector, $\mathbf{D}'$, shall be referred to as the BMQQ matrix vector for the first coordinate, whereas the second vector, $\mathbf{D}''$, will be called BMQQ matrix vector for the second coordinate. Each of these two matrices can be obtained from the other one. In the rest of the paper we will refer both of them as BMQQ matrix vectors (BMQQMV) and we will mostly think for $\mathbf{D}'$. If we want to emphasize that we are dealing with normalized BMQQ, we use nBMQQMV. Next two theorems show that for formula (2) to define a quasigroup operation, matrices $A$ and $B$ must be non-singular.

**Theorem 1.** *Let a BMQQ $(\{0,1\}^n, *)$ be defined by (2). Then the matrices $A$ and $B$ are non-singular.*

**Theorem 2.** *Let $(\{0,1\}^n, *)$ and $(\{0,1\}^n, *')$ be groupoids defined as:*

$$\mathbf{x} * \mathbf{y} = \mathbf{x} + \mathbf{y} + \sum_{i=1}^{n} x_i D_i' \mathbf{y}$$

$$\mathbf{x} *' \mathbf{y} = (A\mathbf{x}) * (B\mathbf{y}) + \mathbf{c},$$

*where $A$ and $B$ are nonsingular $n \times n$ binary matrices and $\mathbf{c}$ is a binary $n$-vector. Then $(\{0,1\}^n, *)$ is a quasigroup iff $(\{0,1\}^n, *')$ is a quasigroup.*

The next theorem is, in fact, an adaptation of Theorem 3 from [14], where additional specifications that were originally presented within the proof have now been explicitly incorporated into the statement. It presents that each BMQQ can be represented using nBMQQ.

**Theorem 3.** *Let a BMQQ $(\{0,1\}^n, *)$ be defined by (2). Then there is a nBMQQ $(\{0,1\}^n, \hat{*})$ and non-singular matrices $A$ and $B$ such that*

$$x * \mathbf{y} = (A\mathbf{x}) \hat{*} (B\mathbf{y}) + \mathbf{c}. \tag{3}$$

*The BMQQMV $\hat{\mathbf{D}} = (\hat{D}_1, \ldots, \hat{D}_n)$ for the quasigroup $(\{0,1\}^n, \hat{*})$ can be expressed in terms of the matrices from (2) as:*

$$(\hat{D}_1, \ldots, \hat{D}_n) = (D_1 B^{-1}, \ldots, D_n B^{-1})(A^{-1})^T. \tag{4}$$

**Definition 2.** *Let BMQQ $(\{0,1\}^n, *)$ be a quasigroup defined by (2). We will call the NBMQQ $(\{0,1\}^n, \hat{*})$ a compatible normal quasigroup for $(\{0,1\}^n, *)$, or shortly **compatible to** $(\{0,1\}^n, *)$, iff they satisfy the relation (3). Accordingly, if $\hat{\mathbf{D}}$ is defined by (4), we will say that it is compatible to $\mathbf{D}$.*

Next two theorems present necessary conditions for an nBMQQMV in a groupoid $(\{0,1\}^n, *)$ to qualify as a quasigroup.

**Theorem 4.** *A vector of binary $n \times n$ matrices $(\hat{D}_1, \ldots, \hat{D}_n)$ is a nBMQQMV for a quasigroup $(\{0,1\}^n, *)$ iff*

$$\forall \alpha_1, \ldots, \alpha_n \in \{0,1\}, |I + \sum_{i=1}^{n} \alpha_i \hat{D}_i| = 1,$$

*where $I$ is the identity $n \times n$ matrix.*

**Theorem 5.** *The vector of binary $n \times n$ matrices $(\hat{D}_1, \ldots, \hat{D}_n)$ represent an nBMQQMV for a quasigroup $(\{0,1\}^n, *)$ iff $\forall \alpha_1, \ldots, \alpha_n \in \{0,1\}$ such that at least one $\alpha_i$ is different then 0, i.e. $\prod_{i=1}^{n}(1 + \alpha_i) = 0$, the vector $\sum_{i=1}^{n} \alpha_i \mathbf{e}_i$ does not belong to the vector space spanned by the column vectors of matrix $\sum_{i=1}^{n} \alpha_i \hat{D}_i$ and does not belong to the vector space spanned by the row vectors of the same matrix $\sum_{i=1}^{n} \alpha_i \hat{D}_i$.*

Following theorem provides a method for constructing a new quasigroup from an existing one.

**Theorem 6.** *Let $(D_1, \ldots, D_n)$ be nBMQQMV and let $A = [a_{ij}]_{n \times n})$ be a non-singular binary matrix. Then the vector $(\hat{D}_1, \ldots, \hat{D}_n)$ defined by*

$$(\hat{D}_1, \ldots, \hat{D}_n) = (A^{-1}D_1A, \ldots, A^{-1}D_nA)A, \tag{5}$$

*is also a nBMQQMV.*

If the matrix vectors nBMQQMV of two nBMQQ quasigroups are connected by the relation (5), then we call them *isotopic relative to matrix representation*.

## 3    Another Way to Express BMQQ

The elements $d_{k,i,j}$ from Definition 1 actually define a three-dimensional cube, i.e. tensor. Note that BMQQMV can be regarded as tensors, but in our specific problem, the order matters, so it is more appropriate to consider them as a vector of matrices rather than as a tensor. In fact, if we arrange the matrices $D'_i$ one behind the other along the X-axis, parallel to the Y-Z plane, they will form a cube in which the matrices at the intersections parallel to the X-Z plane are actually the matrices $D''_i$. In this section, we will analyze the role of the matrices along the third dimension of that cube, namely those that are parallel with the X-Y plane. These matrices serve as a representation of the quasigroup operation, similar as opperation represenatiton in multivariate cryptography [16].

**Definition 3.** *Let $\mathbf{D} = (D_1, \ldots, D_n)$ be nBMQQMV for a given quasigroup $(\{0,1\}^n, *)$. Let $\mathbf{F} = (F_1, \ldots, F_n)$ consists of matrices $F_i$, where the j-th row of the matrix $F_i$ is equal to the i-th row of the matrix $D_i$. Since $F_i$ are parallel with Z-plane will call $\mathbf{F}$ normalized Z BMQQ matrix vector (nZBMQQMV).*

*If $\mathbf{D}$ and $\mathbf{F}$ pertain to the same quasigroup, then we say that $\mathbf{F}$ is corresponding to $\mathbf{D}$ and $\mathbf{D}$ is corresponding to $\mathbf{F}$.*

It's evident that every nBMQQ is uniquely characterized by $\mathbf{F}$. The next two theorems provide a way to represent a quasigroup operation using $\mathbf{F}$.

**Theorem 7.** *Given nBMQQ $(\{0,1\}^n, *)$ with a nZBMQQMV $(F_1, \ldots, F_n)$, the i-th coordinate of the vector $\mathbf{x} * \mathbf{y}$, $(\mathbf{x} * \mathbf{y})_i$, is equal to:*

$$(\mathbf{x} * \mathbf{y})_i = x_i + y_i + c_i + \mathbf{x}^T F_i \mathbf{y} \tag{6}$$

*Proof.* **Proof:** Assume that the corresponding nBMQQMV for the given quasigroup is $(D_1, \ldots, D_n)$. From Definition 1 we have:

$$(\mathbf{x} * \mathbf{y})_i = x_i + y_i + c_i + \sum_{j=1}^{n} x_j (\mathbf{D_{ji}}^T \mathbf{y}),$$

where $\mathbf{D_{ji}}$ is the $i$-th row of the matrix $D_j$, which is the same with the $j$-th row of the matrix $F_i$, $\mathbf{F_{ij}}$. Hence:

$$(\mathbf{x} * \mathbf{y})_i = x_i + y_i + c_i + \left( \sum_{j=1}^{n} x_j \mathbf{D_{ji}}^T \right) \mathbf{y}$$
$$= x_i + y_i + c_i + \left( \sum_{j=1}^{n} x_j \mathbf{F_{ij}}^T \right) \mathbf{y} = x_i + y_i + c_i + \mathbf{x}^T F_i \mathbf{y}.$$

**Theorem 8.** *Let $(\{0,1\}^n, *)$ be a BMQQ defined by (2) and $(\{0,1\}^n, \hat{*})$ be a be its corresponding nBMQQ with nZBMQQMV $(\hat{F}_1, \ldots, \hat{F}_n)$. Then the i-th coordinate of $(\mathbf{x} * \mathbf{y})$ is equal to*

$$(\mathbf{x} * \mathbf{y})_i = (A\mathbf{x})_i + (B\mathbf{y})_i + c_i + \mathbf{x}^T (A^T \hat{F}_i B) \mathbf{y}.$$

*Terms $(A\mathbf{x})_i$ and $(B\mathbf{y})_i$ are the i-th coordinates of vectors $A\mathbf{x}$ and $B\mathbf{y}$ respectively. The vector of binary $n \times n$ matrices $(F_1, \ldots, F_n)$, where $F_i = A^T \hat{F}_i B$, is referred to as an ZBMQQMV for $(\{0,1\}^n, *)$.*

*Proof.* **Proof:** From $(\mathbf{x} * \mathbf{y})_i = (A\mathbf{x} \hat{*} B\mathbf{y})_i$ we have:

$$(\mathbf{x} * \mathbf{y})_i = (A\mathbf{x})_i + (B\mathbf{y})_i + c_i + (A\mathbf{x})^T \hat{F}_i (B\mathbf{y})$$
$$= (A\mathbf{x})_i + (B\mathbf{y})_i + c_i + \mathbf{x}^T (A^T \hat{F}_i B) \mathbf{y}.$$

The proofs of the following two theorems are omitted since Theorem 9 is obvious and follows directly from the previous results given in this chapter and the proof of Theorem 10 is given in [14].

**Theorem 9.** *The vector of binary $n \times n$ matrices $(F_1, \ldots, F_n)$ is nZBMQQMV iff $\forall \alpha_1, \ldots, \alpha_n \in \{0,1\}$, the vector $\sum_{i=1}^{n} \alpha_i \mathbf{e}_i$ is not in the vector spaces generated by the row-vectors and the column-vectors of the matrix $\sum_{i=1}^{n} \alpha_i F_i$.*

**Theorem 10.** *If the vector of binary $n \times n$ matrices $(F_1, \ldots, F_n)$ is nZBMQQMV then the i-th row vector of the matrix $F_i$ is a linear combination of the other row vectors of $F_i$ and the i-th column vector of the matrix $F_i$ is a linear combination of the other column vectors of $F_i$.*

## 4   Comutative BMQQ

In this chapter, we present the form of commutative BMQQ quasigroups.

**Theorem 11.** *The quasigroup* $(\{0,1\}^n, *)$ *defined by (2) is commutative if only if* $A = B$.

*Proof.* **Proof:** Let us assume that the $(\{0,1\}^n, *)$ is a commutative quasigroup. Then $\forall \mathbf{x} \in \{0,1\}^n$, $\mathbf{x} * \mathbf{0} = \mathbf{0} * \mathbf{x}$. From this, it follows that

$$\mathbf{c} + A\mathbf{x} + B\mathbf{0} + \sum_{i=1}^{n} x_i D_i \mathbf{0} = \mathbf{c} + A\mathbf{0} + B\mathbf{x} + \sum_{j=1}^{n} 0 \cdot D_j \mathbf{x}.$$

Hence, $\forall \mathbf{x} \in \{0,1\}^n$, $A\mathbf{x} = B\mathbf{x}$ which is equivalent to $A = B$.

**Theorem 12.** *Let* $(\{0,1\}^n, *)$ *be a NBMQQ with nZBMQQMV* $\mathbf{F} = (F_1, \dots, F_n)$. *Then it is a commutative quasigroup iff* $\forall i = \overline{1, n}$, $F_i = (F_i)^T$.

*Proof.* **Proof:** Let us assume that the quasigroup is commutative. Then for all unite vectors $\mathbf{e}_j$ and $\mathbf{e}_k$ follows that $\mathbf{e}_j * \mathbf{e}_k = \mathbf{e}_k * \mathbf{e}_j$. From (6) we have

$$(\mathbf{e}_j)_i + (\mathbf{e}_k)_i + c_i + (\mathbf{e}_j)^T F_i \mathbf{e}_k = (\mathbf{e}_j)_i + (\mathbf{e}_k)_i + c_i + (\mathbf{e}_k)^T F_i \mathbf{e}_j.$$

This implies that

$$(\mathbf{e}_j)^T F_i \mathbf{e}_k = (\mathbf{e}_k)^T F_i \mathbf{e}_j.$$

But, $(\mathbf{e}_j)^T F_i \mathbf{e}_k$ is the element in the $j$-th row and $k$-th column in the matrix $F_i$, while $(\mathbf{e}_k)^T F_i \mathbf{e}_j$ is the element in the $k$-th row and $j$-th column in the matrix $F_i$. Hence, $F_i = (F_i)^T$.

Let us now assume the opposite, i.e. that $F_i = (F_i)^T$. Then

$$(\mathbf{x} \hat{*} \mathbf{y})_i = x_i + y_i + c_i + (\mathbf{x})^T F_i(\mathbf{y}) = x_i + y_i + c_i + (\mathbf{y})^T (F_i)^T(\mathbf{x})$$
$$= x_i + y_i + c_i + (\mathbf{y})^T F_i(\mathbf{x}) = (\mathbf{y} \hat{*} \mathbf{x})_i.$$

The last theorem of this chapter provides a complete characterization of commutative BMQQs.

**Theorem 13.** *Let* $(\{0,1\}^n, *)$ *be a quasigroup defined by (2) with ZBMQQMV* $\mathbf{F} = (F_1, \dots, F_n)$. *Then* $(\{0,1\}^n, *)$ *is a commutative quasigroup iff* $A = B$ *and* $\forall i = \overline{1, n}$, $F_i = (F_i)^T$.

*Proof.* **Proof:** From Theorem 8 we have that there is a NBLCO-qasigroup $(\{0,1\}^n, \hat{*})$ with nZBMQQMV $(\hat{F}_1, \dots, \hat{F}_n)$ such that the i-th coordinate of $(\mathbf{x} * \mathbf{y})$ is equal to

$$(\mathbf{x} * \mathbf{y})_i = (A\mathbf{x})_i + (B\mathbf{y})_i + c_i + \mathbf{x}^T (A^T \hat{F}_i B)\mathbf{y}.$$

If the quasigroup is commutative, then from pervious theorems $A = B$ and $\forall i = \overline{1, n}, \hat{F}_i = (\hat{F}_i)^T$. Then $\forall i = \overline{1, n}$,

$$F_i = A^T \hat{F}_i A = A^T \hat{F}_i^T A = (A^T \hat{F}_i A)^T = F_i^T.$$

Opposite, if $A = B$ and $\forall i = \overline{1,n}$, $F_i = (F_i)^T$, then

$$\hat{F}_i = (A^{-1})^T F_i A^{-1} = ((A^{-1})^T F_i^T A^{-1})^T = ((A^{-1})^T F_i A^{-1})^T = \hat{F}_i^T.$$

so

$$(\mathbf{x} * \mathbf{y})_i = (A\mathbf{x})_i + (A\mathbf{y})_i + c_i + \mathbf{x}^T (A^T \hat{F}_i A)\mathbf{y}$$
$$= (A\mathbf{y})_i + (A\mathbf{x})_i + c_i + \mathbf{y}^T (A^T \hat{F}_i A)\mathbf{x} = (\mathbf{y} * \mathbf{x})_i.$$

## 5   Associative BMQQ

In this part we provide characterization of associative BMQQ and BMQQ that are commutative semigroups.

**Theorem 14.** *Let* $(\{0,1\}^n, *)$ *be an BMQQ defined by (2) and let* $(\{0,1\}^n, \hat{*})$ *be its compatible NBMQQ. Assume that the nBMQQMVs of* $(\{0,1\}^n, \hat{*})$ *are* $(D_1', \ldots, D_n')$ *and* $(D_1'', \ldots, D_n'')$. *Then* $(\{0,1\}^n, *)$ *is associative, if and only if the following statements are true:*

*i)* $Ac = Bc$. *We will denote this vector with* $\hat{\mathbf{c}} = (\hat{c}_1, \ldots, \hat{c}_n)^T$.
*ii)* $AB = BA$
*iii)* $A + I = \left( \sum_{i=1}^{n} \hat{c}_i D_i'' \right)$ *and* $B + I = \left( \sum_{i=1}^{n} \hat{c}_i D_i' \right)$.
*iv)* $\forall i = \overline{1,n}$, $AD_i' = D_i'A$ *and* $BD_i'' = D_i''B$.
*v)* $\forall \mathbf{x}, \mathbf{y}, \sum_{i=1}^{n} x_i D_i' B\mathbf{y} = \sum_{i=1}^{n} y_i D_i'' A\mathbf{x}$.
*vi)* $\forall i, j = \overline{1,n}$, $D_i' D_j'' = D_j'' D_i'$ .

*Proof.* **Proof:** Note that $\mathbf{0} * \mathbf{0} = \mathbf{c}$, $\mathbf{x} * \mathbf{0} = A\mathbf{x} + \mathbf{c}$ and $\mathbf{0} * \mathbf{x} = B\mathbf{x} + \mathbf{c}$. Assume that the $(\{0,1\}^n, *)$ is associative. Then $\forall \mathbf{x} \in \{0,1\}^n$,

$$\mathbf{0} * (\mathbf{x} * \mathbf{0}) = (\mathbf{0} * \mathbf{x}) * \mathbf{0} \Leftrightarrow \mathbf{0} * (A\mathbf{x} + \mathbf{c}) = (B\mathbf{x} + \mathbf{c}) * \mathbf{0}.$$

$$\Leftrightarrow B(A\mathbf{x} + \mathbf{c}) + \mathbf{c} = A(B\mathbf{x} + \mathbf{c}) + \mathbf{c} \Leftrightarrow BA\mathbf{x} + B\mathbf{c} = AB\mathbf{x} + A\mathbf{c}.$$

If we take $\mathbf{x} = \mathbf{0}$, we prove i). Because $Ac = Bc$, the last expression is equivalent with:

$$BA\mathbf{x} = AB\mathbf{x}.$$

This is true for all $\mathbf{x} \in \{0,1\}^n$, so we have that $AB = BA$, which proves ii).

To prove iii) we use that $(\mathbf{x} * \mathbf{0}) * \mathbf{0} = \mathbf{x} * (\mathbf{0} * \mathbf{0})$ for all $\mathbf{x} \in \{0,1\}^n$. For the left side we have:

$$(\mathbf{x} * \mathbf{0}) * \mathbf{0} = A(\mathbf{x} * \mathbf{0}) + \mathbf{c} = A(A\mathbf{x} + \mathbf{c}) + \mathbf{c} = A^2\mathbf{x} + A\mathbf{c} + \mathbf{c}.$$

The right hand side is reduced to:

$$\mathbf{x} * (\mathbf{0} * \mathbf{0}) = \mathbf{x} * \mathbf{c} = A\mathbf{x} + B\mathbf{c} + \mathbf{c} + \sum_{i=1}^{n} (A\mathbf{x})_i D_i' B\mathbf{c}.$$

If we equate both expressions, having in mind that $Ac = Bc = \hat{c}$, we get:

$$A^2\mathbf{x} = A\mathbf{x} + \sum_{i=1}^{n} (A\mathbf{x})_i D_i' \hat{c}.$$

By moving $A\mathbf{x}$ to the left side and replacing $(A\mathbf{x})_i D_i' \hat{c}$ with $\hat{c}_i D_i'' A\mathbf{x}$, we obtain that $(A + I)A\mathbf{x} = (\sum_{i=1}^{n} \hat{c}_i D_i'')A\mathbf{x}$ holds for all $\mathbf{x} \in \{0, 1\}^n$. This proves the first part of iii). The second part is proved analogously.

To prove iv) we start with $\mathbf{x} * (\mathbf{y} * \mathbf{0}) = (\mathbf{x} * \mathbf{y}) * \mathbf{0}$. Substituting $*$ we obtain:

$$\mathbf{x} * (A\mathbf{y} + \mathbf{c}) = A(\mathbf{x} * \mathbf{y}) + \mathbf{c} \Leftrightarrow$$
$$A\mathbf{x} + B(A\mathbf{y} + \mathbf{c}) + \mathbf{c} + \sum_{i=1}^{n} (A\mathbf{x})_i D_i' B(A\mathbf{y} + \mathbf{c})$$
$$= A(A\mathbf{x} + B\mathbf{y} + \mathbf{c} + \sum_{i=1}^{n} (A\mathbf{x})_i D_i' B\mathbf{y}) + \mathbf{c}.$$

Using i) and ii) the last equality can be reduced to:

$$(A + I)A\mathbf{x} + \sum_{i=1}^{n} (A\mathbf{x})_i D_i' BA\mathbf{y} + \sum_{i=1}^{n} (A\mathbf{x})_i D_i' \hat{c} = A \sum_{i=1}^{n} (A\mathbf{x})_i D_i' B\mathbf{y}.$$

The first and the third term in the last equation are the same, from iii), so since we have binary plus, the last equation is further reduce to:

$$\sum_{i=1}^{n} (A\mathbf{x})_i D_i' BA\mathbf{y} = A \sum_{i=1}^{n} (A\mathbf{x})_i D_i' B\mathbf{y}.$$

Using ii) and the fact that this holds for all $\mathbf{y}$, i.e. all $B\mathbf{y}$, we have:

$$\forall \mathbf{x}, \sum_{i=1}^{n} (A\mathbf{x})_i D_i' A = \sum_{i=1}^{n} (A\mathbf{x})_i A D_i'.$$

Taking $A\mathbf{x} = \mathbf{e}_i$ we get $D_i' A = A D_i'$. This completes the proof of the first part iv). The second part is proved analogously starting from $\mathbf{0} * (\mathbf{x} * \mathbf{y}) = (\mathbf{0} * \mathbf{x}) * \mathbf{y}$.

Since $(\{0, 1\}^n, *)$ is associative, we have that for all $\forall \mathbf{x}, \mathbf{y}, \mathbf{z} \in \{0, 1\}^n$, $\mathbf{x} * (\mathbf{y} * \mathbf{z}) = (\mathbf{x} * \mathbf{y}) * \mathbf{z}$. This is equivalent with:

$$A\mathbf{x} + B\left(A\mathbf{y} + B\mathbf{z} + \mathbf{c} + \sum_{i=1}^{n} (B\mathbf{z})_i D_i'' A\mathbf{y}\right) + \mathbf{c}$$
$$+ \sum_{i=1}^{n} (A\mathbf{x})_i D_i' B\left(A\mathbf{y} + B\mathbf{z} + \mathbf{c} + \sum_{i=1}^{n} (B\mathbf{z})_i D_i'' A\mathbf{y}\right)$$
$$= A\left(A\mathbf{x} + B\mathbf{y} + \mathbf{c} + \sum_{i=1}^{n} (A\mathbf{x})_i D_i' B\mathbf{y}\right) + B\mathbf{z} + \mathbf{c}$$
$$+ \sum_{i=1}^{n} (B\mathbf{z})_i D_i'' A\left(A\mathbf{x} + B\mathbf{y} + \mathbf{c} + \sum_{i=1}^{n} (A\mathbf{x})_i D_i' B\mathbf{y}\right)$$

Using i), ii),iii) and iv) we obtain

$$\sum_{i=1}^{n}(A\mathbf{x})_i(D_i'B)B\mathbf{z} + \sum_{i=1}^{n}(A\mathbf{x})_i D_i'B\left(\sum_{i=1}^{n}(B\mathbf{z})_i D_i''A\mathbf{y}\right)$$
$$= \sum_{i=1}^{n}(B\mathbf{z})_i(D_i''A)A\mathbf{x} + \sum_{i=1}^{n}(B\mathbf{z})_i D_i''A\left(\sum_{i=1}^{n}(A\mathbf{x})_i D_i'B\mathbf{y}\right)$$

Taking $\mathbf{y} = \mathbf{0}$ we obtain following equality which is equivalent with v).

$$\sum_{i=1}^{n}(A\mathbf{x})_i(D_i'B)B\mathbf{z} = \sum_{i=1}^{n}(B\mathbf{z})_i(D_i''A)A\mathbf{x}.$$

Since the last equality is true for associative quasigroup we have:

$$\sum_{i=1}^{n}\sum_{j=1}^{n}(A\mathbf{x})_i(B\mathbf{z})_j D_i'BD_j''A\mathbf{y} = \sum_{j=1}^{n}\sum_{i=1}^{n}(B\mathbf{z})_j(A\mathbf{x})_i D_j''AD_i'B\mathbf{y}$$

Using ii) and iv) the last equality can be rewritten as:

$$\sum_{i=1}^{n}\sum_{j=1}^{n}(A\mathbf{x})_i(B\mathbf{z})_j D_i'D_j''BA\mathbf{y} = \sum_{j=1}^{n}\sum_{i=1}^{n}(B\mathbf{z})_j(A\mathbf{x})_i D_j''D_i'BA\mathbf{y}.$$

This holds for all $\mathbf{x}$ and $\mathbf{y}$, so it must be $D_j''D_i' = D_i'D_j''$.

To prove the opposite, i.e. that i) to iv) imply associativity, we can start from $\mathbf{x} * (\mathbf{y} * \mathbf{z}) = (\mathbf{x} * \mathbf{y}) * \mathbf{z}$ and reduced it to $0 = 0$.

The next theorem reduces the requirements of Theorem 14.

**Theorem 15.** *Let* $(\{0,1\}^n, *)$ *be an BMQQ defined by (2) and let* $(\{0,1\}^n, \hat{*})$ *be its compatible NBMQQ. Assume that the nBMQQMVs of* $(\{0,1\}^n, \hat{*})$ *are* $(D_1', \ldots, D_n')$ *and* $(D_1'', \ldots, D_n'')$. *Then* $(\{0,1\}^n, *)$ *is associative if and only if the following statements are true:*

*i)* $A\mathbf{c} = B\mathbf{c} = \hat{\mathbf{c}};$

*ii)* $A + E = \sum_{i=1}^{n}\hat{\mathbf{c}}_i D_i''$ *and* $B + E = \sum_{i=1}^{n}\hat{\mathbf{c}}_i D_i';$

*iii)* $\forall i,j = \overline{1,n}, D_i'D_j'' = D_j''D_i'.$

*Proof.* **Proof:** We need to show that the requirements ii), iv) and v) from Theorem 14 are redundant, i.e. can be obtained from the other requirements in the Theorem 14.

First we will proof that $AB = BA$ follows from i), ii) and iii) from this theorem. We have:

$$AB = \left(E + \sum_{i=1}^{n} \widehat{\mathbf{c}}_i D_i''\right)\left(E + \sum_{j=1}^{n} \widehat{\mathbf{c}}_j D_j'\right)$$

$$= E + \sum_{i=1}^{n} \widehat{\mathbf{c}}_i D_i'' + \sum_{j=1}^{n} \widehat{\mathbf{c}}_j D_j' + \left(\sum_{i=1}^{n} \widehat{\mathbf{c}}_i D_i''\right)\left(\sum_{j=1}^{n} \widehat{\mathbf{c}}_j D_j'\right)$$

$$= E + \sum_{i=1}^{n} \widehat{\mathbf{c}}_i D_i'' + \sum_{j=1}^{n} \widehat{\mathbf{c}}_j D_j' + \sum_{i=1}^{n}\sum_{j=1}^{n} \widehat{\mathbf{c}}_j \widehat{\mathbf{c}}_i D_i'' D_j'.$$

Similarly, we can obtain that

$$BA = E + \sum_{i=1}^{n} \widehat{\mathbf{c}}_i D_i' + \sum_{j=1}^{n} \widehat{\mathbf{c}}_j D_j'' + \sum_{i=1}^{n}\sum_{j=1}^{n} \widehat{\mathbf{c}}_j \widehat{\mathbf{c}}_i D_i' D_j''.$$

From $D_i' D_j'' = D_j'' D_i'$ it follows that the two expressions are equal.

Next we are proving that $\forall i = \overline{1,n}, D_i' A = A D_i'$.

$$D_j' A = D_j'\left(E + \sum_{i=1}^{n} \widehat{\mathbf{c}}_i D_i''\right) = D_j' + \sum_{i=1}^{n} \widehat{\mathbf{c}}_i D_j' D_i''.$$

From $D_i' D_j'' = D_j'' D_i'$ this is equal to

$$D_j' A = D_j' + \sum_{i=1}^{n} \widehat{\mathbf{c}}_i D_i'' D_j' = \left(E + \sum_{i=1}^{n} \widehat{\mathbf{c}}_i D_i''\right) D_j' = A D_j'.$$

The prove for $\forall i = \overline{1,n}, D_i'' B = B D_i''$ is analogues to this one.

At the end we give the proof that $\forall \mathbf{x}, \mathbf{y}, \sum_{i=1}^{n} x_i D_i' B\mathbf{y} = \sum_{i=1}^{n} y_i D_i'' A\mathbf{x}$. Replacing $B$ with $E + \sum_{j=1}^{n} \widehat{c}_j D_j'$ we obtain:

$$\sum_{i=1}^{n} x_i D_i' B\mathbf{y} = \sum_{i=1}^{n} x_i D_i'\left(E + \sum_{j=1}^{n} \widehat{c}_j D_j'\right)\mathbf{y}$$

$$= \sum_{i=1}^{n} x_i D_i' \mathbf{y} + \sum_{i=1}^{n} x_i D_i' \sum_{j=1}^{n} \widehat{c}_j D_j' \mathbf{y}$$

$$= \sum_{i=1}^{n} x_i D_i' \mathbf{y} + \sum_{i=1}^{n}\sum_{j=1}^{n} x_i y_j D_i' D_j'' \widehat{c}$$

Similarly,

$$\sum_{i=1}^{n} y_i D_i'' A\mathbf{x} = \sum_{i=1}^{n} y_i D_i'' \mathbf{x} + \sum_{i=1}^{n}\sum_{j=1}^{n} y_i x_j D_i'' D_j' \widehat{c}.$$

The first terms of this two expressions are equal from the definition of the matrices $D_i^{''}$ and $D_j^{'}$, and the second terms are equal because $D_i^{''} D_j^{'} = D_i^{'} D_j^{'}$.

This last Theorem provides a procedure for generating other semigroups from a given semigroup $(\{0,1\}^n, *)$. In fact, we may choose a constant vector $\widehat{\mathbf{c}}$ and construct the matrices $A$ and $B$ as in Theorem 15 $ii)$. Taking $\mathbf{c} = A^{-1}\widehat{\mathbf{c}}$ ensures that the quasigroup defined with $A\mathbf{x} * B\mathbf{y} + \mathbf{c}$ is a semigroup.

Next Theorem follows directly from the Theorem 11 and Theorem 15.

**Theorem 16.** *Let $(\{0,1\}^n, *)$ be an BMQQ defined by (2) and let $(\{0,1\}^n, \hat{*})$ be its compatible nBMQQ. Assume that the nBMQQMVs of $(\{0,1\}^n, *')$ are $(D_1^{'}, \ldots, D_n^{'})$ and $(D_1^{''}, \ldots, D_n^{''})$. Then $(\{0,1\}^n, *)$ is commutative semigroup, if and only if the following statements are true:*

*i) $A = B$*
*ii) $\forall i = \overline{1,n}, D_i^{'} = D_i^{''} = D_i$*
*iii) $A + E = \sum_{i=1}^{n} (A\mathbf{c})_i D_i$*
*iv) $\forall i, j = \overline{1,n}, D_i D_j = D_j D_i$ .*

## 6  Conclusion and Future Work

Results presented in this paper are natural extension of the our previous work on Bilinear Multivariate Quadratic Quasigroups (BMQQs). Building upon our prior research, we have delved into a novel perspective on the representation of quasigroups. In this study, we specifically focus on providing a comprehensive framework for two essential classes of quasigroups: associative and commutative quasigroups, which represent a solid foundation for further exploration and open avenues for future research in the study of quasigroups. Our findings have significant implications and can be applied in various domains. They offer valuable insights for designing algorithms to generate quasigroups and for verifying whether a given quasigroup satisfies the properties of BMQQ. Additionally, our research contributes to the analysis of algorithm performance in coding, encryption, and other applications that utilize quasigroups as a fundamental tool.

**Acknowledgements.** This work was partially financed by the Faculty of Computer Science and Engineering at the Ss. Cyril and Methodius University in Skopje.

## References

1. Allsop, J., Wanless, I.M.: Row-Hamiltonian Latin squares and Falconer varieties (2022). https://arxiv.org/abs/2211.13826
2. Gill, M.J., Wanless, I.M.: Perfect 1-factorisations of K16. Bull. Aust. Math. Soc. **101**, 177–185 (2020)

3. Ding, J., Petzoldt, A.: Current state of multivariate cryptography. IEEE Secur. Priv. **15**(4), 28–36 (2017)
4. Chen, Y., Gligoroski, D., Knapskog, S.J.: On a special class of multivariate quadratic quasigroups (MQQs). J. Math. Cryptol. **7**(8), 111–141 (2013)
5. Chauhan, D., Gupta, I., Verm, R.: Quasigroups and their applications in cryptography. Cryptologia **45**(3) (2021)
6. Drápal, A., Wanless, I.M.: Isomorphisms of quadratic quasigroups (2022). https://arxiv.org/abs/2211.09472
7. Galatenko, A.V., Nosov, V.A., Pankratiev, A.E.: Generation of multivariate quadratic quasigroups by proper families of Boolean function. J. Math. Sci. **262**, 630–641 (2022)
8. Zhang, Y., Zhang, H.: An algorithm for judging and generating multivariate quadratic quasigroups over Galois fields. Springerplus **5**(1), 1845 (2016)
9. Gligoroski, D., Markovski, S., Knapskog, S.J.: A public key block cipher based on multivariate quadratic quasigroups (2008). https://arxiv.org/abs/0808.0247
10. Faugère, J.C., Gligoroski, D., Perret, L., Samardjiska, S., Thomae, E.: A Polynomial-Time Key-Recovery Attack on MQQ Cryptosystems. In: Katz, J. (eds.) Public-Key Cryptography – PKC 2015. PKC 2015. LNCS, vol. 9020, pp. 150–174. Springer, Berlin, Heidelberg (2015). https://doi.org/10.1007/978-3-662-46447-2_7
11. Mohamed, M.S.E., Ding, J., Buchmann, J., Werner, F.: Algebraic attack on the MQQ public key cryptosystem. In: Garay, J.A., Miyaji, A., Otsuka, A. (eds.) Cryptology and Network Security. CANS 2009. LNCS, vol. 5888, pp. 392–401. Springer, Berlin, Heidelberg (2009). https://doi.org/10.1007/978-3-642-10433-6_26
12. Siljanoska, M., Mihova, M., Markovski, S.: Matrix presentation of quasigroups of order 4. In: Proceedings of the 10th Conference for Informatics and Information Technologies (CIIT 2013), pp. 192–196 (2014)
13. Mihova, M., Siljanoska, M., Markovski, S.: Tracing bit differences in strings transformed by linear quasigroups of order 4. In: Proceedings of the 9th Conference for Informatics and Information Technology (CIIT 2012), Bitola, Macedonia, pp. 229–233 (2012)
14. Mihova, M., Stojanova, A.: On bilinear quasigroups of order 2n. In: Proceedings of the ICT Innovations (2019)
15. Samardjiska, S., Chen, Y., Gligoroski, D.: Construction of multivariate quadratic quasigroups (MQQs). In: Proceedings of the 7th International Conference on Information Assurance and Security (IAS) in Arbitrary Galois Fields (2011)
16. Huang, Y.J., Liu, F.H., Yang, B.Y.: Public-key cryptography from new multivariate quadratic assumptions. In: Fischlin, M., Buchmann, J., Manulis, M. (eds.) Public Key Cryptography – PKC 2012. PKC 2012. LNCS, vol. 7293, pp. 190–205. Springer, Berlin, Heidelberg (2012). https://doi.org/10.1007/978-3-642-30057-8_12

# Introducing Probabilities in Networks of Polarized Splicing Processors

Victor Mitrana[1,2(✉)] and Mihaela Păun[2,3]

[1] Department of Information Systems, Polytechnic University of Madrid,
Calle Alan Turing s/n, 28031 Madrid, Spain
victor.mitrana@upm.es

[2] National Institute of Research and Development for Biological Sciences,
296 Independenței Bd. District 6, 060031 Bucharest, Romania
mihaela.paun@incdsb.ro

[3] Faculty of Administration and Business, University of Bucharest,
Bucharest, Romania

**Abstract.** Motivated by the need of reducing the huge amount of data navigating simultaneously through a network of polarized splicing processors, we look to the possibility of introducing probabilities which theoretically could decrease this amount, at a price of some loss of certainty. We imagined two possible situations regarding the splicing step: to associate either fixed or dynamically computed probabilities with splicing rules in every node. Similarly to the splicing step, two situations could be considered for the communication step depending on the way the probabilities are associated: statically or dynamically. We believe that this new feature together with the communication protocol based on polarization might facilitate software simulations or hardware implementations.

**Keywords:** DNA computing · Splicing processor · Polarization · Network of polarized splicing processors · Probability

## 1 Introduction

Networks of polarized splicing processors (NSP for short) belong to the family of bio-inspired computational models sharing several characteristics: (i) the data is encoded in strings, (ii) the operation which is the computational source is abstracted from a biological phenomenon, that of splicing, and (iii) the architecture is a distributed one and the computation is highly parallel. Many of these bio-inspired computational models are computationally powerful, being able to simulate Turing machines, but they have been used to theoretically solve hard problems see, e.g., [20,22]. Networks of splicing processors have been introduced in [10] and investigated in a series of subsequent publications, see, e.g., [13,16],

This work was performed through the Core Program within the National Research, Development and Innovation Plan 2022-2027, carried out with the support of MRID, project no. 23020101(SIA-PRO), contract no 7N/2022, and project no. 23020301 (SAFE-MAPS), contract no 7N/2022.

for a discussion about this topic. Informally, a network of splicing processors is an undirected graph whose vertices are hosts for processors running the splicing operation on data encoded as strings.

A splicing processor is designed to run an operation called *splicing* that is inspired from the recombination of double stranded DNA molecules by means of restriction enzymes and DNA ligase [9]. Figure 1 illustrates this operation, where two DNA sequences are cut by restriction enzymes and then recombined by the DNA ligase.

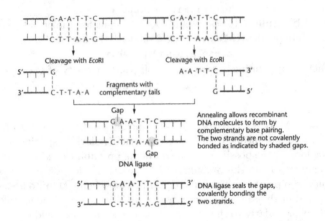

**Fig. 1.** Splicing operation.

Formally, this process is defined as a string rewriting operation as shown in [8]: each restriction enzyme has been approximated by a splicing rule that defines the DNA sequences recognized by that enzyme, as well as the restriction sites. The computation in a network of splicing processors consists in splicing and communication steps which alternate with each other. In each splicing step, in a highly parallel manner, each processor changes the strings it contains according to its associated sets of splicing rules.

In the earlier models, the communication has been regulated by filters associated with nodes that allow/block strings from going out nodes and/or entering nodes. Each filter has been defined by some context-conditions that allow or forbid the presence of some symbols in the current string. In the communication step, all the strings that pass the corresponding filters are simultaneously interchanged between the connected nodes. This model is related to other distributed computational models based on splicing like the model proposed in [4] and [18]. Another related model is a language generating model whose investigation started in [17] and continued in [14,19]. The differences between all these models are discussed in [12]. Theoretical solutions to hard computational problems with networks of splicing processors have been proposed in [11].

In [2] the protocol of communication was changed such that it is not anymore regulated by filters but by a sort of compatibility between nodes and data, called polarity. This model is called network of polarized splicing processors

(NPSP). We informally explain this protocol. Each processor is either negatively, or positively charged, but also without any charge (neutral charge). This is formally done by assigning a sign in the set $\{-, 0, +\}$ to every processor. Unlike the charge of a processor, the charge of data is dynamically computed as the algebraic sum of the charge values associated with each symbol of the string. The protocol has been extensively discussed in [16] for different variants of networks of cell-like processors. Thus, one may metaphorically say that the strings are electrically polarized. Consequently, the protocol of communication is naturally regulated by the compatibility of the polarity of a node and that of data, namely each string will migrate to a node of the same polarity as it has.

Aspects such as stochastic, fuzzy or rough set concepts that are often identifiable in the behavior, dynamics, self-learning and evolutionary paths in biological systems have not been studied at all for any variant of network of splicing processors until now. The work [3] is an exhaustive and convincing work dealing with various possibilities offered by probabilistic/stochastic approaches when modeling biochemical phenomena. We discuss here the possibility to associate probabilities to the splicing rules and strings. This study is motivated by the difficulties of implementing the NSP. Indeed, due to its huge parallelism and nondeterminism, it is difficult to store such amount of data. What is our intention here is to introduce probabilities in order to decrease the amount of data during the computation in networks of polarized splicing processor at the price of losing some certainty of the output. The output is obtained now with some probability because the strings with a very low probability, which are estimated to have no essential role in the final result, are ignored.

We mainly consider two possible ways of introducing probabilities in NPSP in a similar way to that used for networks of polarized evolutionary processors [1]. We give here just an informal description. At the beginning of any computation, the input string appears with probability 1 in the input node. All the axioms have the same probability, namely 1, at every step. This means that before a splicing step is to start, all axioms associated with every node are newly introduced. We now discuss the two possibilities of introducing probabilities associated to the splicing rules. One possibility is to associate a predefined probability with each splicing rule existing in a node. Now for any string, having a probability of being present in that node, which is involved in a splicing step with an axiom, one computes a preliminary probability of the two strings obtained. This is done by taking into account the probability of the string (that of an axiom is always 1) and the probability of the splicing rule that has been applied. At the end of a splicing step, the probability of every string of a node is calculated taking into account its preliminary probabilities. A similar calculation is to be done at the end of a communication step considering that all nodes are equally probable to receive the same string that is compatible to them. Finally, the computation halts at some moment, if the sum of all probabilities of the strings existing in a specific node, called the halting node, is higher than a given value. This is called, the probability of the halting node to be nonempty. If this probability is at least some value $\chi$, then the input string is validated with probability $\chi$. We also propose another condition for validating an input string when the computation halts. This condition could be more useful because it gives the possibility to invalidate the input string in a simpler manner.

Another possible way of introducing the probabilities associated with the splicing rules is as follows. In every splicing step, for each node, we consider that all the splicing rules *which can be applied* to a pair of strings are equally probable, hence the probability of each rule is not predefined anymore but it may change before each splicing step. It is worth mentioning that in the same splicing step, the probability of the same rule might differ depending on the pair of string which is applied to. The communication protocol may be modified to follow a similar dynamical approach. We may interpret this approach by considering that the localization of a string in the network is given by the probability of that string at any computational step. Returning to biochemistry, our source of inspiration, this way of introducing probabilities might be seen as a rough estimation/expectation of the place of different molecules and/or their quantity in a biological system.

In the final section we discuss some open problems and further direction of research.

## 2    Preliminaries

For any finite set $A$, $card(A)$ denotes the cardinality of $A$ and for a string $w$, $|w|$ denotes the length of $w$. The smallest alphabet $W$ such that $w \in W^*$ is denoted by $alph(w)$ and the empty string is denoted by $\lambda$.

A *multiset* over a set $X$ (see, e.g., [1]) is a mapping $\mu : X \longrightarrow \mathbf{N}$; $\mu(a)$ expresses the number of copies of $a \in X$ in the multiset $\mu$. The *support* of $\mu$ is defined by $supp(\mu) = \{a \in X \mid \mu(a) \neq 0\}$. A multiset $\mu$ is said to be *finite* if $supp(\mu)$ is finite. The *empty multiset* over $X$ is denoted by $\Lambda$, that is $\Lambda(a) = 0$ for all $a \in X$. The *cardinality* of a multiset $\mu$ as above is $\#(\mu) = \sum_{a \in X} \mu(a)$; moreover, we set $\#_Y(\mu) = \sum_{a \in Y} \mu(a)$ for any subset $Y$ of $X$. As a general rule, the elements of a set are listed as usual between { and } while the elements of a finite multiset are listed between [ and ]. For instance, the multiset $\mu$ over the set of binary strings such that $\mu(00) = 2$, $\mu(111) = 3$, and $\mu(x) = 0$ for all binary strings $x$ different than 00 and 111 can be written as $[00, 111, 00, 111, 111]$.

For two multisets $\mu_1, \mu_2$ over the same set $X$ we define

- the *multiset addition* $\mu_1 + \mu_2$, defined by $(\mu_1 + \mu_2)(a) = \mu_1(a) + \mu_2(a)$ for all $a \in X$;
- the *multiset difference* $\mu_1 - \mu_2$, defined by $(\mu_1 - \mu_2)(a) = \mu_1(a) - \mu_2(a)$ for each $a \in X$, provided that $\mu_2 \sqsubseteq \mu_1$;

By [15], an *(integer) valuation* is a homomorphism from $V^*$ to $\mathbf{Z}$, which associates an integer with each string over an alphabet $V$. For a string $w$, this integer is computed as the algebraic sum of the values associated to the letters of $w$. We denote by $abs(k)$ the absolute value of the integer $k$.

We now recall the definition of the splicing operation following [8]. A *splicing rule* over a finite alphabet $V$ is a quadruple of strings of the form $[(u_1, u_2); (v_1, v_2)]$ such that $u_1$, $u_2$, $v_1$, and $v_2$ are in $V^*$. For a splicing rule $r = [(u_1, u_2); (v_1, v_2)]$ and for $x, y, w, z \in V^*$, we say that $r$ produces $z$ from $x$ and $y$ (denoted by $(x, y) \vdash_r z$) if there exist some $x_1, x_2, y_1, y_2 \in V^*$ such that $x = x_1 u_1 u_2 x_2$, $y = y_1 v_1 v_2 y_2$, and $z = x_1 u_1 v_2 y_2$. For two strings $x$, $y$, and a splicing rule $r$, we write

$$\sigma_r(x, y) = \{z \mid (x, y) \vdash_r z\} \cup \{z \mid (y, x) \vdash_r z\}.$$

For two set of strings $L_1, L_2$ over $V$ and a set of splicing rules $R$ we define

$$\sigma_R(L_1, L_2) = \bigcup_{\substack{x \in L_1, \\ y \in L_2, \\ r \in R}} \sigma_r(x, y).$$

A *polarized splicing processor* over $V$ is a triple $(S, A, \pi)$ where

- $S$ a finite set of splicing rules over $V$,
- $A$ a finite set of auxiliary strings over $V$,
- $\pi \in \{-, +, 0\}$ is the polarization of the node (negatively or positively charged, or neutral, respectively).

A *network of polarized splicing processors* (NPSP for short) is a construct

$$\Gamma = (V, U, \langle, \rangle, G, \mathcal{N}, \varphi, \underline{In}, \underline{Halt}),$$

where

- $U$ is the network alphabet and $V \subseteq U$ is the input alphabet.
- $\langle, \rangle \in U \backslash V$ are two special symbols.
- $G = (X_G, E_G)$ is an undirected graph with nodes $X_G$ and edges $E_G$.
- $\mathcal{N}$ is a mapping which associates with each node $x \in X_G$ the polarized splicing processor over $U$, $\mathcal{N}(x) = (S_x, A_x, \pi_x)$.
- $\varphi$ is a valuation of $U^*$ in $\mathbf{Z}$.
- $\underline{In}$ and $\underline{Halt}$ are the *input* and the *halting* nodes, respectively.

The graph $G$ is called the *underlying graph* of the network. We say that $card(X_G)$ is the size of $\Gamma$.

A *configuration* of an NPSP is defined by the sets of strings (without their multiplicities) existing in every node of the network before (or after) a computational step. Formally, this is defined as a function $C : X_G \to 2^{U^*}$, such that $C^k(x)$ is the set of strings existing in the node $x$ at the $k^{th}$ step of a computation. For a string $w \in V^*$ the initial configuration of $\Gamma$ on $w$ is defined by $C_0^{(w)}(x_{\underline{In}}) = \{\langle w \rangle\}$ and $C_0^{(w)}(x) = \emptyset$ for all other $x \in X_G$. We consider that no auxiliary string appears in any configuration.

Each splicing step, as well as a communication one, changes the current configuration into another one. In a splicing step, each set $C(x)$ of the current configuration $C$ is changed in accordance with the splicing rules of the node $x$ which are applied to all possible pairs formed by a string in $C(x)$ and the other in $A_x$. In a formal way, the configuration $C$ is changed into the configuration $C'$, and we write this as $C \Rightarrow C'$, iff for all $x \in X_G$

$$C'(x) = \sigma_{S_x}(C(x), A_x).$$

In a communication step, each node $x$ does in parallel the following:

- it sends copies of all its strings to all the nodes connected to $x$ and keeps a local copy of the strings having the same polarity as itself;
- it receives a copy of each string sent by any node connected with itself, providing that the string has the same polarity as that of $x$.

A short explanation is needed here. In nature, each electrically charged particle/molecule migrates towards the opposite pole. We have preferred, by two reasons, that each node migrates to a node having the same polarity. One reason is that this approach gives us the possibility to consider that also the neutral strings can migrate. The second reason is for simplicity.

Formally, the configuration $C$ is changed into the configuration $C'$ by a *communication step*, written as $C \vdash C'$, iff

$$C'(x) = (C(x) \backslash \{w \in C(x) \mid sign(\varphi(w)) \neq \pi_x\}) \cup$$
$$\bigcup_{\{x,y\} \in E_G} (\{w \in C(y) \mid sign(\varphi(w)) = \pi_x\}),$$

for all $x \in X_G$. We have denoted by $sign(m)$ the function which returns the sign of $m$, that is $+, 0, -$. It is important to mention that, according to the definition, if a string existing in the node $x$ has a different polarity than that of $x$, it is expelled from $x$. If a string has been expelled and it cannot enter any node connected to the node from where it was expelled, the string is lost.

Let $\Gamma$ be an NPSP, the computation of $\Gamma$ on the input string $w \in V^*$ is a sequence of configurations $C_0^{(w)}, C_1^{(w)}, C_2^{(w)}, \ldots$, where $C_0^{(w)}$ is the initial configuration of $\Gamma$ on $w$, $C_{2i}^{(w)} \Longrightarrow C_{2i+1}^{(w)}$ and $C_{2i+1}^{(w)} \vdash C_{2i+2}^{(w)}$, for all $i \geq 0$. As one can see, the configurations are changed by alternative steps. We can also note that each configuration is changed in a deterministic way.

A computation as above *halts* if one of the next two conditions is satisfied:

(i) the computation reaches a configuration in which the halting node <u>Halt</u> is non-empty,
(ii) the computation cannot continue anymore.

Given an NPSP $\Gamma$ that halts on every input, and an input string $w$, we say that $\Gamma$ validate $w$ if the computation of $\Gamma$ on $w$ halts by the first condition, namely <u>Halt</u> is non-empty.

# 3  Introducing Probabilities

In the first part of this section, we discuss two different ways of associating probabilities with the splicing rules. The simplest way could be to consider that the splicing rules are equally probable, that is they have associated the same probability. Returning to biochemistry, which was our source of inspiration, this situation is not very common either "in vivo" or "in vitro". Consequently, a more realistic approach would be to associate with each rule a predefined probability, and this is the first way of introducing probabilities at the level of splicing rules we are to discuss in the sequel. This could be interpreted in the way that some reactions are inhibited while others are more active, depending on the actual conditions of the environment. Along the same lines, we may suppose that all the inhibited reactions have the same probability, namely zero. It follows that, the probability associated with the other rules that might be active are dynamically computed with respect to the strings existing in a node at some moment. This possibility will also be analysed in what follows.

A *polarized splicing processor over V, with rules having predefined probabilities* is a quadruple $(S, A, \pi, p)$, where:

- $(S, A, \pi, p)$ is a splicing processor over $V$;
- $p$ is a probability distribution on $M$, $p : S \longrightarrow (0, 1]$ such that $\sum_{r \in S} p(r) = 1$.

A *network of polarized splicing processors with rules having predefined probabilities* (NPSPP for short) is defined as an NPSP with the only difference that $\mathcal{N}$ associates now with each node $x$ the polarized splicing processor with rules having predefined probabilities, $\mathcal{N}(x) = (S_x, A_x, \pi_x, p_x)$.

The definition of a *configuration* of an NPSPP $\Gamma$ is changed a bit, being defined by the mapping $C : X_G \longrightarrow 2^{U^* \times (0,1]}$ which now associates a set of pairs (string, nonzero value in the unit interval) with every node of the graph. Each such pair may be understood as the probability of the string to be present in that node. Given a string $w \in U^*$, the initial configuration of $\Gamma$ on $w$ is defined by $C_0^{(w)}(\underline{In}) = \{(w, 1)\}$ and $C_0^{(w)}(x) = \emptyset$ for all $x \in X_G \backslash \{\underline{In}\}$.

A configuration can change exactly like in an NPSP, namely by a *splicing step* and a *communication step* which alternate each other. We need some preparation before giving the formal definition of a splicing step in some node $x$. We first compute the multiset $\mu_{C(x)}$ with Algorithm 1.

---

**Algorithm 1. Input:** $C(x)$.
$\quad\quad\quad\quad$ **Output:** $\mu_{C(x)}$.

---

1: $\mu_{C(x)} := \Lambda$;
2: **for** $((z, q) \in C(x)) \& (u \in A_x) \& (r \in S_x))$ **do**
3: $\quad$ **for** $z' \in \sigma_r(z, u)$ **do**
4: $\quad\quad$ $q' := q \cdot p(r)$;
5: $\quad\quad$ $\mu_{C(x)} := \mu_{C(x)} + [(z', q')]$;
6: $\quad$ **end for**
7: **end for**

---

As soon as we have computed $\mu_{C(x)}$, we can now compute the mapping $\rho_x : U^* \longrightarrow [0,1]$ by the Algorithm 2.

---

**Algorithm 2. Input: $z$ and $\mu_{C(x)}$.**
          **Output: $\rho_x(z)$.**

---

1: $\rho_x(z) := 0$;
2: **while** (exists ($q \in (0,1]$) and $(z,q) \in supp(\mu_{C(x)})$) **do**
3:     Choose $(z,q) \in supp(\mu_{C(x)})$;
4:     $\mu_{C(x)} := \mu_{C(x)} - [(z,q)]$;
5:     $\rho_x(z) := \rho_x(z) + q$;
6: **end while**

---

Now, we can define the change of a configuration, that is the configuration $C$ is changed into $C'$ by a splicing step, written as $C \Longrightarrow C'$, iff

$$C'(x) = \{(z, \rho_x(z)) \mid \rho_x(z) \neq 0\} \text{ for all } x \in X_G.$$

Every string that appear in the next configuration associated with $x$ is obtained exactly like in a splicing step in an NPSP. Instead of considering more copies of this string, we associate with it a probability which is computed in two steps by calculating the probability of two independent events. In the first step, a preliminary positive value in the unitary interval is computed for each new string as follows. As the probability of each axiom to be present in a node is always 1, we compute the product between the probability of the string which is to be spliced and the probability of the splicing rule that was applied. Then, the new string appears in the next configuration with a probability which is the sum of the values, computed in the first step, of all copies of that string which have been obtained in that splicing step.

In order to define the polarized splicing processor with rules having dynamical probabilities, we need a short preparation. Let us consider a node $x$ and a current configuration $C_k^{(w)}(x)$. For each pair $(z,q)$ that belongs to $C_k^{(w)}(x)$, we define

$$R_k^{(w)}(x,z) = \{r \in S_x \mid r \text{ can be applied to } (z,y), \text{ for some } y \in A_x, y \neq z\}.$$

Now the probability associated with the rule $r \in S_x$ with respect to the string $z$ in the next splicing step is defined as

$$p_k^{(w)}(r) = \begin{cases} \frac{1}{card(R_k^{(w)}(x,z))}, & \text{if } r \in R_k^{(w)}(x,z), \\ 0, & \text{otherwise.} \end{cases}$$

A *network of polarized splicing processors with rules having dynamic probabilities* (NPSDP for short) is defined as an NPSPP with the only difference that each node $x$ is a polarized splicing processor with rules having dynamically computed probabilities as above.

The splicing step is defined exactly as above with the difference that in the first step, when a preliminary positive value in the unitary interval is computed for each new string, the probability used is that dynamically computed as above with respect to the spliced string (different than the axiom).

In the same way as above, we may consider two possible ways of associating probabilities to the nodes. The first one is again given by a static assignation of probabilities. Since the communication step depends only on the probabilities associated with the nodes, we give below the more interesting case when these probabilities are dynamically computed.

The way of changing a configuration by a communication step is actually the same to that of a communication step in a network of probabilistic evolutionary processors [1]. For the sake of self-containment, we recall it here. As in the case of the splicing step, we need to preprocess the following numbers.

For a node $x \in X_G$ and $q \in \{-1, 0, 1\}$ we define

$$deg_q(x) = card\{y \in X_G \mid \{x, y\} \in E_G, \pi_y = q\} + \begin{cases} 1, & \text{if } \pi_x = q, \\ 0, & \text{if } \pi_x \neq q. \end{cases}$$

Informally, $deg_q(x)$ is the number of nodes adjacent to $x$ that have the polarity $q$. Note that this number counts $x$ itself, provided that $x$ has the polarity $q$.

In the same preprocessing phase, we compute the following set for every configuration $C$ and node $x$:

$$Out(C(x)) = \{(z, q') \mid (z, q) \in C(x),$$
$$q' = \frac{q}{deg_{sign(\varphi(z))}(x)}, deg_{sign(\varphi(z))}(x) \neq 0\}.$$

As we have done for a splicing step, we need to calculate the multiset $\mu_{C(x)}$ for a communicating step by the next algorithm:

---

**Algorithm 3. Input:** $C(x)$.
**Output:** $\mu_{C(x)}$.

---

1: $\mu_{C(x)} := [(z, t) \mid (z, t) \in Out(C(x)), sign(\varphi(z)) = \pi_x]$;
2: **for** $y \in X_G$ such that $\{x, y\} \in E_G$ **do**
3:    $\mu_{C(x)} := \mu_{C(x)} + [(z, t) \mid (z, t) \in Out(C(y))$;
4:    $sign(\varphi(z)) = \pi_x]$;
5: **end for**

---

After computing the multisets $\mu_{C(x)}$, $x \in X_G$, we proceed with the computation of the mappings $\rho_x$ exactly like in the case of a splicing step discussed above. We now formally define the communication step. The configuration $C$ is changed into the configuration $C'$ by a communication step, written as $C \vdash C'$, iff

$$C'(x) = \{(z, \rho_x(z)) \mid \rho_x(z) \neq 0\} \text{ for all } x \in X_G.$$

Again $\rho$ is computed exactly as in the case of a splicing step.

Some explanations about our approach. Let us consider that a string in a node $x$ at a given step has the same polarity to that of some nodes connected to $x$. Our proposal is that the string migrates with equal probability to all these nodes. Remember that even the node $x$ retains a copy of this string, provided that the string has the same polarity to that of $x$. On the other hand, it makes sense to consider that all strings having polarities different than that of $x$ must be expelled with probability 1. Clearly, as in the case of an NPSP, each expelled string from a node $x$, that cannot enter node connected to $x$ by the reason of different polarity is lost.

The computation of $\Gamma$ on the input string $w \in V^*$ is defined exactly as for the networks without probabilities.

Finally, we discuss the halting condition in any of the variants considered above. Obviously, a possible condition for halting a computation would be the same to that for NSPP. At the first sight, this does not make too much sense as the probabilities calculated during the computation play no role. However, as we see in the sequel, this still might be a possibility- Another possibility is to define the halting condition with respect to a cut-off point $\chi \in (0,1]$, that is, the computation halts if there exists a configuration in which the sum of probabilities of all the strings existing in the halting node is at least $\chi$. Formally, there exists $k$ such that

$$\sum_{(z,q)\in C_k^{(w)}(\underline{Halt})} q \geq \chi.$$

Another possibility to halt a computation with a cut-off point $\chi \in (0,1]$ could be if there exists $k$ such that

$$\max\{p \mid (z,p) \in C_k^{(w)}(\underline{Halt}), \text{ for some } z\} \geq \chi,$$

holds. Remember that, we said that an input string is validated by an NSPP, that halts on every input, if the computation on that input halts by a non-empty halting node.

Having probabilities, we may refine this condition. One possibility is to validate a string $w$ with cut-off point $\mu$ if there exists a halting computation (without cut-off point) $C_0^{(w)}, C_1^{(w)}, C_2^{(w)}, \ldots, C_k^{(w)}$ such that

$$\sum_{(z,q)\in C_k^{(w)}(\underline{Halt})} q \geq \rho.$$

An analogous condition could be imposed for a halting computation with cut-off point. Clearly, we must have have $\chi \leq \mu$.

## 4    Conclusion and Final Remarks

We have proposed a few possible ways of introducing probabilities in the networks of polarized splicing processors. We don't pretend that we have exhausted all the possibilities, but our ways are especially suggested by biochemistry, our source of inspiration for the initial model and its variants. For example, we have considered the uniform distribution for associating dynamical probabilities with the splicing rules and nodes. However, other probabilistic distributions, such as the Poisson distribution, the Bernoulli distribution or the binomial distribution might be considered.

There were reported several software implementations in JAVA for networks of evolutionary and splicing processors, see, e.g., [5,21] as well as simulations using massively parallel platforms for multi-core computers, clusters of computers and cloud resource [6,7]. Along the same lines, it is worth noting that a simulation of probabilistic networks of polarized evolutionary processors has been proposed in [23]. We intend to investigate the feasibility of implementing these probabilistic variants in a forthcoming work.

# References

1. Arroyo, F., Gómez Canaval, S., Mitrana, V., Păun, M., Sánchez-Couso, J.R.: Towards probabilistic networks of polarized evolutionary processors. In: International Conference on High Performance Computing & Simulation, HPCS 2018, pp. 764–771, IEEE (2018)
2. Bordihn, H., Mitrana, V., Păun, A., Păun, M.: Networks of polarized splicing processors. In: Martín-Vide, C., Neruda, R., Vega-Rodríguez, M.A. (eds.) TPNC 2017. LNCS, vol. 10687, pp. 165–177. Springer, Cham (2017). https://doi.org/10.1007/978-3-319-71069-3_13
3. Bower, J.M., Bolouri, H.: Computational Modeling of Genetic and Biochemical Networks. MIT Press, Cambridge (2001)
4. Csuhaj-Varjú, E., Kari, L., Păun, G.: Test tube distributed systems based on splicing. Comput. AI **15**, 211–232 (1996)
5. Diaz, M.A., Mingo, de L.F., Gömez Blas, N., Castellanos, J.: Implementation of massive parallel networks of evolutionary processors (MPNEP): 3-colorability problem. In: Krasnogor, N., Nicosia, G., Pavone, M., Pelta, D. (eds.) Nature Inspired Cooperative Strategies for Optimization (NICSO 2007). Studies in Computational Intelligence, vol. 129, pp. 399–408. Springer, Heidelberg (2008). https://doi.org/10.1007/978-3-540-78987-1_36
6. Canaval, S.G., de la Puente, A.O., González, P.O.: Distributed simulation of neps based on-demand cloud elastic computation. In: Rojas, I., Joya, G., Catala, A. (eds.) IWANN 2015. LNCS, vol. 9094, pp. 40–54. Springer, Cham (2015). https://doi.org/10.1007/978-3-319-19258-1_4
7. Gómez Canaval, S., Ordozgoiti Rubio, B., Mozo, A.: NPEPE: massive natural computing engine for optimally solving NP-complete problems in big data scenarios. In: Morzy, T., Valduriez, P., Bellatreche, L. (eds.) ADBIS 2015. CCIS, vol. 539, pp. 207–217. Springer, Cham (2015). https://doi.org/10.1007/978-3-319-23201-0_23
8. Head, T.: Formal language theory and DNA: an analysis of the generative capacity of specific recombinant behaviors. Bull. Math. Biol. **49**(6), 737–759 (1987)
9. Head, T., Păun, G., Pixton, D.: Language theory and molecular genetics: generative mechanisms suggested by DNA recombination. In: Rozenberg, G., Salomaa, A. (eds.) Handbook of Formal Languages, pp. 295–360. Springer, Heidelberg (1997). https://doi.org/10.1007/978-3-662-07675-0_7

10. Manea, F., Martín-Vide, C., Mitrana, V.: Accepting networks of splicing processors. In: Cooper, S.B., Löwe, B., Torenvliet, L. (eds.) CiE 2005. LNCS, vol. 3526, pp. 300–309. Springer, Heidelberg (2005). https://doi.org/10.1007/11494645_38

11. Manea, F., Martín-Vide, C., Mitrana, V.: All NP-problems can be solved in polynomial time by accepting networks of splicing processors of constant size. In: Mao, C., Yokomori, T. (eds.) DNA 2006. LNCS, vol. 4287, pp. 47–57. Springer, Heidelberg (2006). https://doi.org/10.1007/11925903_4

12. Manea, F., Martín-Vide, C., Mitrana, V.: Accepting networks of splicing processors: complexity results. Theoret. Comput. Sci. **371**, 72–82 (2007)

13. Manea, F., Martín-Vide, C., Mitrana, V.: Accepting networks of evolutionary word and picture processors: A survey. In: Martín-Vide, C. (ed.) Scientific Applications of Language Methods, World Scientific / Imperial College Press, pp. 525–560 (2010)

14. Margenstern, M., Rogozhin, Y.: Time-varying D distributed H systems of degree 1 generate all recursively enumerable languages. In: Words, Semigroups, and Transductions, pp. 329–340, World Scientific Publishing, Singapore (2001)

15. Martín-Vide, C., Mitrana, V.: P systems with valuations. In: Antoniou, I., Calude, C.S., Dinneen, M.J. (eds.) Unconventional Models of Computation, UMC'2K, Discrete Mathematics and Theoretical Computer Science, pp. 154–166. Springer, London (2001). https://doi.org/10.1007/978-1-4471-0313-4_13

16. Mitrana, V.: Polarization: a new communication protocol in networks of bioinspired processors. J. Membrane Comp. **1**, 1–17 (2019)

17. Păun, G.: DNA computing: distributed splicing systems. In: Mycielski, J., Rozenberg, G., Salomaa, A. (eds.) Structures in Logic and Computer Science. LNCS, vol. 1261, pp. 353–370. Springer, Heidelberg (1997). https://doi.org/10.1007/3-540-63246-8_22

18. Păun, G.: Distributed architectures in DNA computing based on splicing: limiting the size of components. In: Calude, C., Casti, J., Dinneen, M.J. (eds.) Unconventional Models of Computation, pp. 323–335. Springer-Verlag, Berlin (1998)

19. Păun, A.: On time-varying H systems. Bull. EATCS **67**, 157–164 (1999)

20. Păun, Gh., Rozenberg, G., Salomaa, A.: DNA Computing: New Computing Paradigms. Springer, Heidelberg (1998)

21. Rosal, del E., Nñez, R., Ortega. A.: MapReduce: simplified data processing on large clusters., Int. J. of Computers, Communica- tions and Control III, 480–485 (2008)

22. Rozenberg, G., Bäck, T., Kok, J.N. (eds.): Handbook of Natural Computing. Springer, Heidelberg (2012)

23. Sánchez Martin, J.A., Arroyo, F.: Simulating probabilistic networks of polarized evolutionary processors. In: Rudas, I.J., Csirik, J., Toro, C., Botzheim, J., Howlett, R.J., Jain, L.C. (eds) Knowledge-Based and Intelligent Information & Engineering Systems, KES 2019, vol. 159, pp. 1421–1430, Procedia Computer Science. Elsevier (2019)

# Author Index

M. Mihova and M. Jovanov (Eds.): ICT Innovations 2023, CCIS 1991, p. 263, 2024.
https://doi.org/10.1007/978-3-031-54321-0

Printed in the United States
by Baker & Taylor Publisher Services

Printed in the United States
by Baker & Taylor Publisher Services